Co-product Feeds

Animal feeds from the food and drinks industries

Robin Crawshaw

Nottingham
University Press

First published by Nottingham University Press

This reissued original edition published 2023 by 5m Books Ltd
www.5mbooks.com

British Library Cataloguing in Publication Data Co-Product Feeds:
Crawshaw, R.

ISBN 9781789182743

Typeset by Nottingham University Press, Nottingham
Digitally printed and bound by PODWW, UK

Foreword

Events have shown that animal feeds and the materials used in them can have an unfortunate and dramatic impact on the safety of food for human consumption unless particular care is taken by all concerned. The feed industry and its suppliers have therefore taken heed of the potential consequences by stepping up quality assurance in ways described in Robin Crawshaw's book. Moreover, in recognition of the role of animal feed in the food chain, many aspects of the composition and labelling of feed are now the responsibility of the Food Standards Agency.

The Agency puts the consumer first. Besides the obvious need to safeguard consumers' health, it is also important that consumers have the necessary information with which to make informed choices on what they buy and eat. This is equally true of farmers and other purchasers of materials to be used in animal feed. They have to know that feed materials are safe, wholesome and provide the necessary nutrition for the animals' growth.

This book supplies valuable insights and explanation about a wide range of feed materials which derive from the production of food and drink. There is continued and heightened interest, in the European Parliament and elsewhere, about what animals are fed, with particular emphasis on feed labelling and the listing of feed materials in EC law. So, besides those directly involved in the feed and farming industries, legislators will also benefit from the information provided here.

Bill Knock
Head of Animal Feed Division
Food Standards Agency

About the author

Robin Crawshaw
B.Sc, M.Sc, CBiol., MIBiol, R.Nutr.

Robin Crawshaw was educated at Mirfield Grammar School in Yorkshire and at the University College of North Wales, where he graduated in Biochemistry and Soil Science. Mr Crawshaw worked as an advisory nutrition chemist with ADAS and its forerunner NAAS, where he developed a special interest in silage production and utilisation. He was awarded an MSc for a thesis on approaches to the evaluation of silage additives. More recently, his interests have widened considerably to encompass many of the feed materials that become available as a result of processing by the food and drinks industries. It is the experience gained in this later period that has provided the basis for this book. Mr Crawshaw has provided technical support on animal feed matters to MAFF, the Food Standards Agency and the EU, and has served on two "Mini-Groups" established by the European Commission to advise its Committee of Experts. His services have also been provided outside Europe, in Pakistan, Australia and Argentina. For four years he held the position of Agricultural Attaché to the British High Commission in New Zealand. In conjunction with his wife, he now runs an independent feed consultancy business from his home in Cambridgeshire.

Contents

Acknowledgements

The author wishes to acknowledge the help of many people in the preparation of this book. Such undertakings inevitably take the writer outside his own experience, and it is with gratitude that I record my particular thanks to the following individuals and organisations who have helped to supply expert advice in some of those areas.

Acknowledgement must first be made to the Brewing, Food and Beverage Industry Suppliers Association (BFBi), and specifically to its Brewers' Grains and Co-products Committee. When the idea for this book was first presented to this group, it was their ready acceptance of its potential value that provided the initial stimulus. Their subsequent sponsorship, and promotion of the idea to other potential sponsors, helped to make this project a reality. During the last ten years, this Association has provided an effective lobby that has defended the position of brewers' grains and other co-product feeds. It has persuaded Ministers that such feeds should not be labelled as wastes, it has reassured both animal health officials and consumer organisations with regard to feed safety, and it has promoted the nutritional benefits of these materials to some of the toughest judges – the farmer customers. I trust that BFBi members will consider that this book represents a useful addition to the basic message – that these feeds represent safe, nutritious and cost-effective nutrient sources.

I am personally indebted to Mr R T Pass who has been an enthusiastic supporter of this project since its inception, and who has taken the trouble to read and provide comments on the text. I have been encouraged by his support and appreciative of his informed comments.

I should also like to thank the following organisations that have given permission to use previously unpublished information: ADAS, Amylum Group, James & Son (GM) Ltd, United Distillers and Vintners.

Many others have provided me with information, and I have valued this and the opportunity it has provided to discuss various issues with them. However, it should be noted that the comments made in this book represent the author's interpretation of the facts. The following list is a comprehensive guide to the individuals and organisations that have contributed to these discussions.

ADAS: Prof. D I Givens
Agri-Food Market Analysis: Dr D McQueen
Assured Combinable Crops Registrar, Ms Dawn Hall
Alcohols Ltd: Mr J Landers,
Allied Bakeries: Mr K Tomblin
Amylum Group: Ir B Mys, and A van Houte
Amylum UK: Mr D Ward
APV Anhydro, Copenhagen: Ms S Arndal and Mr N Steven
Badminton Horse Feeds: Ms Sue Proctor
Banbury Agriculture: Mr R White
Bass Technical Centre: Mr J MacDonald
BFBi: Dr P G E Bradfield and Mr M J Rayner
BLRA: Mr M Spillane
Brewing Research International: Mr G Freeman
Mr M Bridewell, farmer
British Nutrition Foundation: Ms Claire MacEvilly
British Potato Council: Messrs. P Bicknell and R Burrow
British Soft Drinks Association Ltd: Dr R Hargitt
British Sugar: Messrs. P M Bee, P J Jarvis, K M A Sanigar, J B Smith and S R Todd
H P Bulmers: Ms Loraine Boddington and Messrs R Peers and G Thomas
Campden and Chorleywood Research Station: Dr A Aldrick, Mr T Hutton and Ms Kim Little
Cargill UK: Messrs. S Arundel, J Biggs, G Butler, M Hardy, C Shepherd and Mrs Carol A Wilson
Cerestar UK Ltd: Mr G Fletcher and Miss Jo Stapleton
Combination of Rothes Distillers Ltd: Mr J Paton and Ms Agnes Peters
DETR Waste Policy Division: Messrs M Hughes and J MacIntyre
EFSIS Certification: Mr G Fairweather
Federation of Bakers: Mr J White
Food Standards Agency: Mr W D Knock, Dr R Smith
Bob Forsyth Trading: Mr R Forsyth
Freedom Food Ltd: Miss Nina Crump
Frito Lay: Mr E Robinson
Geoff Pattimore Transport: Mr G Pattimore
Glenmorangie plc: Mr W Lumsden
Gin & Vodka Association: Mr E Atkinson and Ms Tricia Crighton
Greenall Group Ltd: Mr E Hughes
Greenvale Foods: Mr S Ward
Hillcrest: Mr I Reivers

James & Son (Grain Merchants) Ltd: Messrs. P Featherstone, R W Hall and P V Sparks
J R Transport: Mr J Rosewell
Loxton Transport: Mr W Phippen
McCain Foods GB Ltd: Dr P J Harkett and Mr K Wilmot
KW Scotland: Mr P Hill
Kanco: Mr K Anderson,
MAFF: Ms Alison Bromley
Dr C May, formerly Chief Scientist, *Citrus Colloids Ltd*
Mid-Norfolk Produce: Mr C Smedley
Muir UK: Mr D Clifford
Muntons plc: Mr A Auger
Nabim: Mr D Testa
Nestlé UK Ltd: Mr J Rostron
NRM Ltd: Mrs Christine Collins
North British Distillery Co Ltd: Ms Diane O'Connor
Novartis Consumer Health: Mr P Addison
Orchard House Foods: Mr M A Cockerill, Mr A Morizzo and Mrs Claire Streit
Potato Business World: Mr A Brice
Potato Processors Association: Mr R Harris
Quest International: Mr I Horwill and Dr A I Maynard
Rhodia Ltd: Mr J Winwood
Roquette UK: Mr S Grainger
SugaRich: Mr P Latham
Sunjuice: Mr J Hawkins
SWA: Ms Jill Bennett
Taymix Transport Ltd: Mr C D R Taylor
Trident Feeds: Dr M W Witt and Mr A Jackson
UKASTA: Mr B Hill
United Molasses: Mr J Higginbotham
UM Feeds Marketing: Mr C Hammond
United Distillers & Vintners: Messrs R T Pass and D Stewart
Charles Wells Ltd: Mr G McFarlane
Weston Research Laboratories: Dr D J Wallington
Wheyfeeds: Mr S Bridge and Ms Liz Shilton

Robin Crawshaw
RC Feed Research

Sponsors

I am indebted to the member companies of BFBi who so readily put up funds to support this effort. Thanks to this initial support, further contributions were also obtained and the full list of sponsors is shown below. These companies wish to be associated with the publication of this book, and the company logos are printed on the following page.

Bass Brewers Ltd
I M Cowe & Co Ltd
Carlsberg-Tetley Brewing Ltd
Cressy's Grains Ltd
W Freeman & Sons Ltd
Guinness UDV
Interbrew UK Ltd
James & Son (Grain Merchants) Ltd
KW Alternative Feeds
McCain Foods (GB) Ltd
Malt Distillers Association of Scotland
Natural Resources Management Ltd
Scottish Courage Ltd
Trident Feeds
UM Feeds Marketing
Wheyfeeds Ltd

Robin Crawshaw
RC Feed Research

BFBi

CRESSY'S GRAINS LIMITED
ESTABLISHED 1945

COWE agriculture

KW Alternative Feeds

JAMES & SON

GUINNESSUDV

um Feeds Marketing

nrm laboratories

Bass Brewers

CARLSBERG-TETLEY

McCain

William Freeman & Sons Limited

Interbrew UK Ltd

SCOTTISH COURAGE
— LIMITED —

THE MALT DISTILLERS ASSOCIATION OF SCOTLAND

TRIDENT FEEDS

WF Wheyfeed Ltd.
Nottingham

Glossary of Terms and Abbreviations

Some of the terms used in this book may not be familiar to all readers. The following list defines them as they are used in the text.

ABM Assured British Meat

ABTA Allied Brewery Traders Association

ACCS Assured Combinable Crops Scheme

ADAS Agricultural Development and Advisory Service

ADF Acid detergent fibre – a laboratory measure of fibre content (cellulose and lignin)

ADIN Acid detergent insoluble nitrogen – a measure of indigestible protein

ARC Agricultural Research Council

BFBI Brewing, Food and Beverage Industry Suppliers Association

BLRA Brewers and Licensed Retailers Association

BOD Biochemical oxygen demand – a measure of pollution potential

BSDA British Soft Drinks Association

BSE Bovine spongiform encephalopathy – a brain disease of cattle, believed to be transmitted in infected feed.

BSF Badminton Speciality Feeds

CEDAR Centre for Dairy Research, University of Reading

CIRF Corn Industries Research Foundation Inc.

CSL Corn steep liquor – juice obtained by steeping maize grain

CCSL Concentrated corn steep liquor – evaporated CSL

CMS Condensed molasses solids / solubles – concentrated liquor remaining after fermentation and distillation of sugar in molasses

COMA Committee on Medical Aspects of Food

DE Digestible energy - commonly used in the UK as the standard measure of the energy value of feeds for pigs

DM Dry matter – all the constituents of a feed other than water

DMD Dry matter digestibility – the (apparent) proportion of a feed that is not passed out in the faeces

DMSBF Dried molassed sugar beet feed – the principal co-product from sugar beet extraction, comprising a blend of the pulp with molasses

DNA De-oxyribose nucleic acid – the code of genetic information carried by plant and animal cells

DOH Department of Health

DOMD Digestible organic matter in the dry matter – a measure of the value of a feed to an animal expressed as a proportion of the dry matter content

DUP Digestible undegraded protein – the amount of feed protein that is absorbed by ruminant animals beyond the rumen

EC European Community

EFSIS European Food Safety and Inspection Service

EPA United States Environmental Protection Agency

ESW Evaporated spent wash – a concentrated form of the liquid fraction remaining after distillation of alcohol

EU European Union

FABBL Farm Assured British Beef and Lamb

FAO Food and Agriculture Organisation of the United Nations

FOB Federation of Bakers

FSA Food Standards Agency

GDG Grain distillers grains – a dried blend of co-products from a grain distillery (undefined by cereal type)

GDG-M Maize distillers grains - a dried blend of maize co-products from a grain distillery

GDG-W Wheat distillers grains - a dried blend of wheat co-products from a grain distillery

GE Gross energy – the energy value of a feed determined by burning it in oxygen

GM Genetic modification / genetically modified – a laboratory technique for implanting desirable characteristics (in agricultural crops) by means of an alteration of the genetic code.

GNS Grain neutral spirit – alcohol used in the production of gin and vodka

GVA Gin & Vodka Association

HACCP	Hazard analysis of critical control points – a logical approach to the assessment of risk
IOB	Institute of Brewing
ISO	International Standards Organisation
LPF	Liquid potato feed – the outer layers of potato separated after subjection of washed tubers to a short period of pressurised steam
MAFF	Ministry of Agriculture, Fisheries & Food
MDA	Malt Distillers Association
MDG	Malt distillers grains – usually refers to the moist, solid co-product from a malt distillery
ME	Metabolisable energy - commonly used in the UK as the standard measure of the energy value of feeds for ruminant animals
MGF	Maize gluten feed – the principal co-product of the wet-milling of maize grain, comprising a blend of maize fibre and concentrated corn steep liquor
MGM	Maize gluten meal – a high protein co-product of maize fractionation
MJ	Megajoule – a standard unit of energy
MMF	Masham Micronised Feeds
MRL	Maximum residue limit – refers to the maximum amount of pesticide that would be expected on a harvested crop grown in accordance with good farming practice
MRP	Malt residual pellets – a pelleted blend of the feed co-products produced by the maltster
MSBF	Molassed sugar beet feed – only available in dried form, this is a short-hand form of DMSBF
Nabim	National Association of British and Irish Millers
NACM	National Association of Cider Makers
NACNE	National Advisory Committee on Nutrition Education
NCD	Neutral cellulase digestibility – a laboratory enzyme procedure for assessing the digestibility of a feed
NCGD	Neutral cellulase and gamanase digestibility – a modified version of NCD, using enzymes with a wider substrate re-activity
NDC	National Dairy Council

NDF Neutral detergent fibre - a laboratory measure of fibre content (hemicellulose, cellulose and lignin)

NDFAS National Dairy Farm Assurance Scheme

NFE Nitrogen-free extract – a chemical term that describes the undefined fraction of a feed after accounting for the protein, oil, fibre, ash and water contents

OMD Organic matter digestibility - a measure of the value of a feed to an animal expressed as a proportion of the organic matter content

PAS Pot ale syrup – a concentrated form of the liquor remaining in a malt distillery still after alcohol distillation

PCR Polymerize chain reaction – a laboratory procedure for the analysis of DNA to identify genetically modified material

PPF Prime potato feed – a moist blend of potato co-products, comprising the steam-detached peel and the centrifuged cell contents that are released when peeled potatoes are cut into chips

PPP Prime potato puree – a concentrated form of the cell contents released when peeled potatoes are chipped

QA Quality assurance

RRI Rowett Research Institute

RSPCA Royal Society for the Prevention of Cruelty to Animals

SAC Scottish Agricultural College

SPII Scottish Pig Industry Initiative

SQBLA Scottish Quality Beef and Lamb

SQC Scottish Quality Cereals Scheme

SWA Scotch Whisky Association

TASCC Trade Assurance Scheme for the Storage and Haulage of Combinable Crops

UDV United Distillers and Vintners

UFAS UKASTA Feed Assurance Scheme

UKAS United Kingdom Accreditation Service

UKASTA United Kingdom Agricultural Supply Trade Association

WPSA World Poultry Science Association

WQL Welsh Quality Lamb

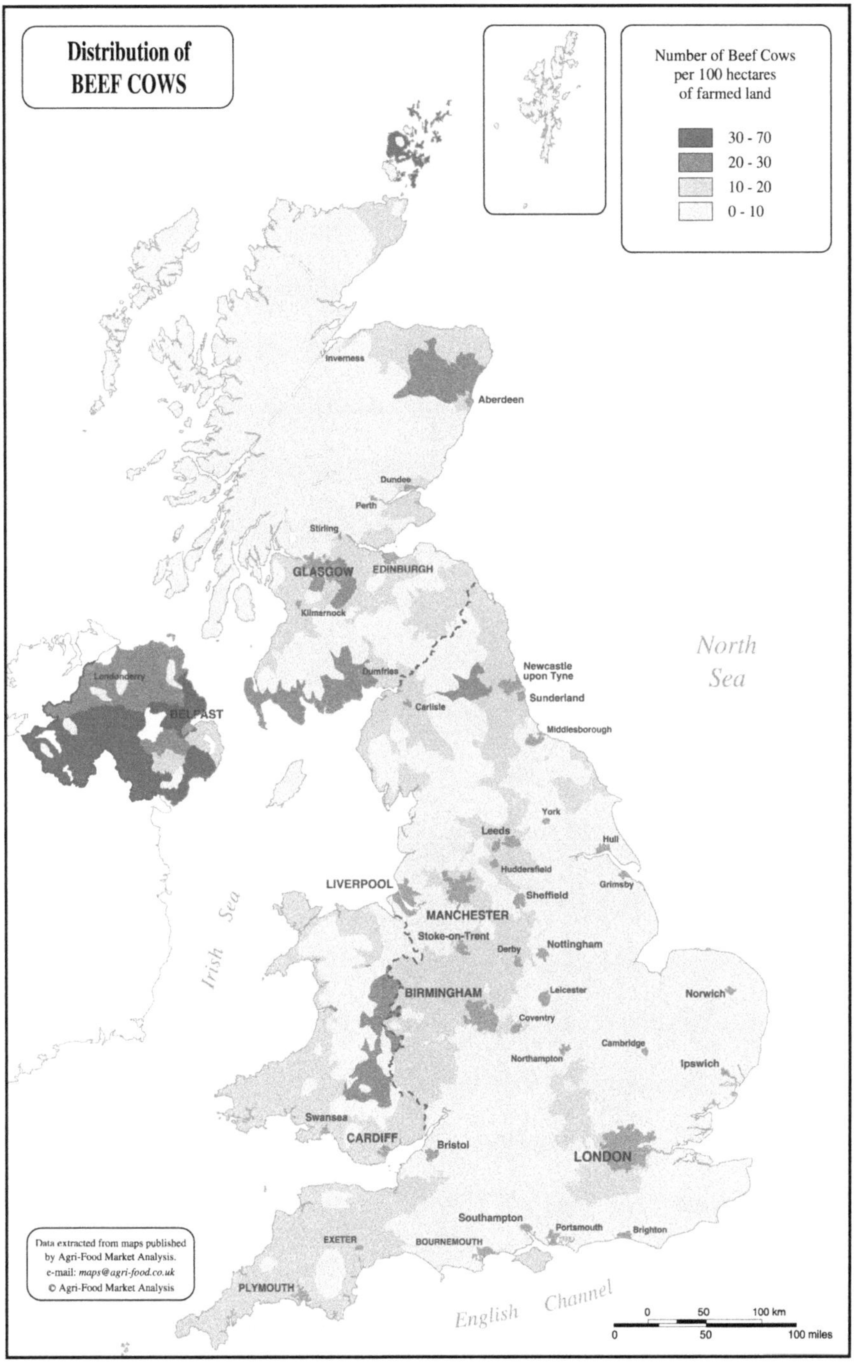
Distribution of
BEEF COWS
Number of Beef Cows
per 100 hectares
of farmed land
30 - 70
20 - 30
10 - 20
0 - 10
Inverness
Aberdeen
Dundee
Perth
Stirling
GLASGOW
EDINBURGH
Kilmarnock
Londonderry
BELFAST
Dumfries
Carlisle
Newcastle
upon Tyne
Sunderland
Middlesborough
North
Sea
York
Leeds
Hull
Huddersfield
Grimsby
LIVERPOOL
Sheffield
MANCHESTER
Stoke-on-Trent
Derby
Nottingham
Irish Sea
BIRMINGHAM
Leicester
Norwich
Coventry
Cambridge
Northampton
Ipswich
Swansea
CARDIFF
Bristol
LONDON
Southampton
Portsmouth
Brighton
EXETER
BOURNEMOUTH
PLYMOUTH
English Channel
0 50 100 km
0 50 100 miles
Data extracted from maps published
by Agri-Food Market Analysis.
e-mail: maps@agri-food.co.uk
© Agri-Food Market Analysis

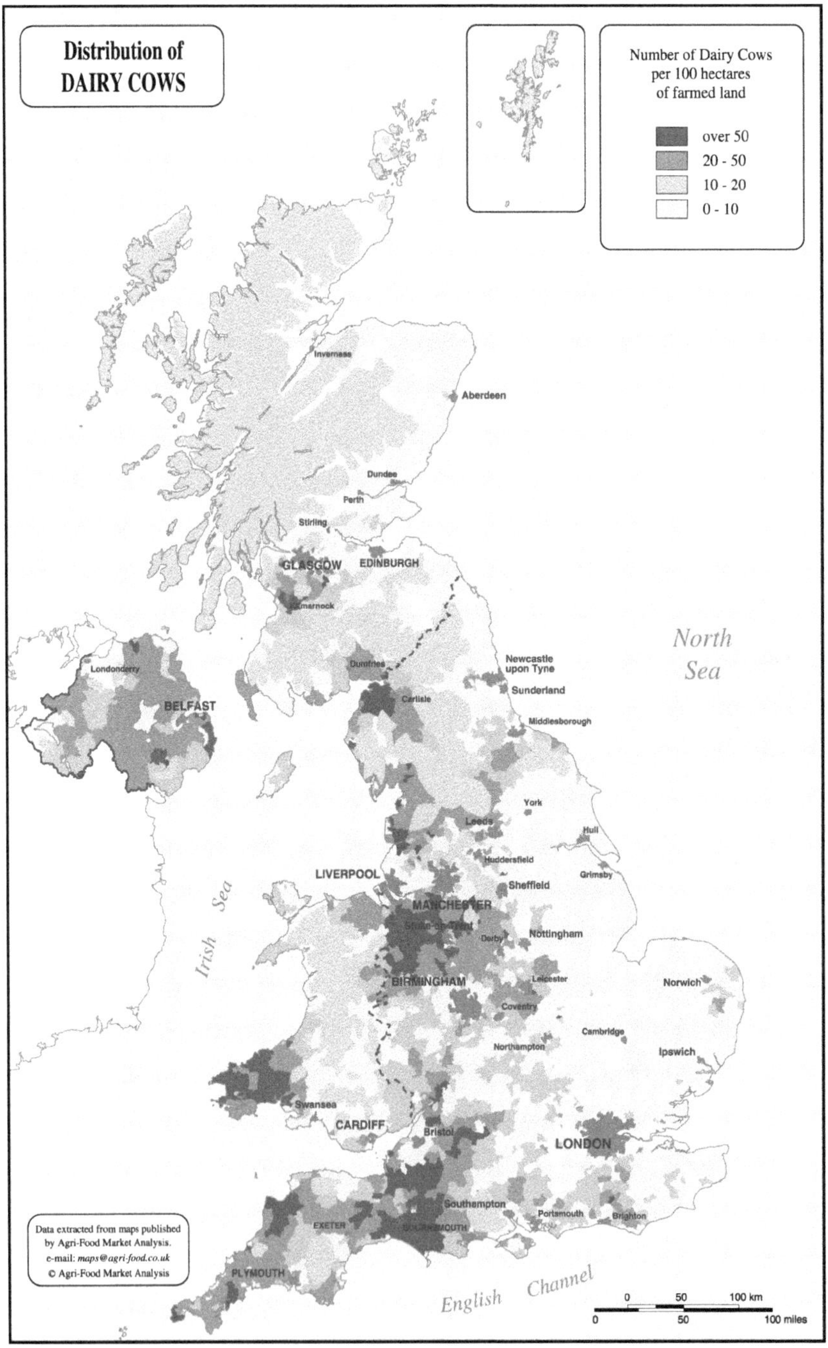
Distribution of
DAIRY COWS
Number of Dairy Cows
per 100 hectares
of farmed land
over 50
20 - 50
10 - 20
0 - 10
Inverness
Aberdeen
Dundee
Perth
Stirling
GLASGOW
EDINBURGH
Dumfries
Londonderry
BELFAST
Carlisle
Newcastle
upon Tyne
Sunderland
Middlesborough
North
Sea
York
Leeds
Hull
Huddersfield
Grimsby
LIVERPOOL
Sheffield
Irish Sea
MANCHESTER
Derby
Nottingham
BIRMINGHAM
Leicester
Coventry
Norwich
Cambridge
Northampton
Ipswich
Swansea
CARDIFF
Bristol
LONDON
Southampton
Portsmouth
Brighton
EXETER
PLYMOUTH
English Channel
0
50
100 km
0
50
100 miles
Data extracted from maps published
by Agri-Food Market Analysis.
e-mail: maps@agri-food.co.uk
© Agri-Food Market Analysis

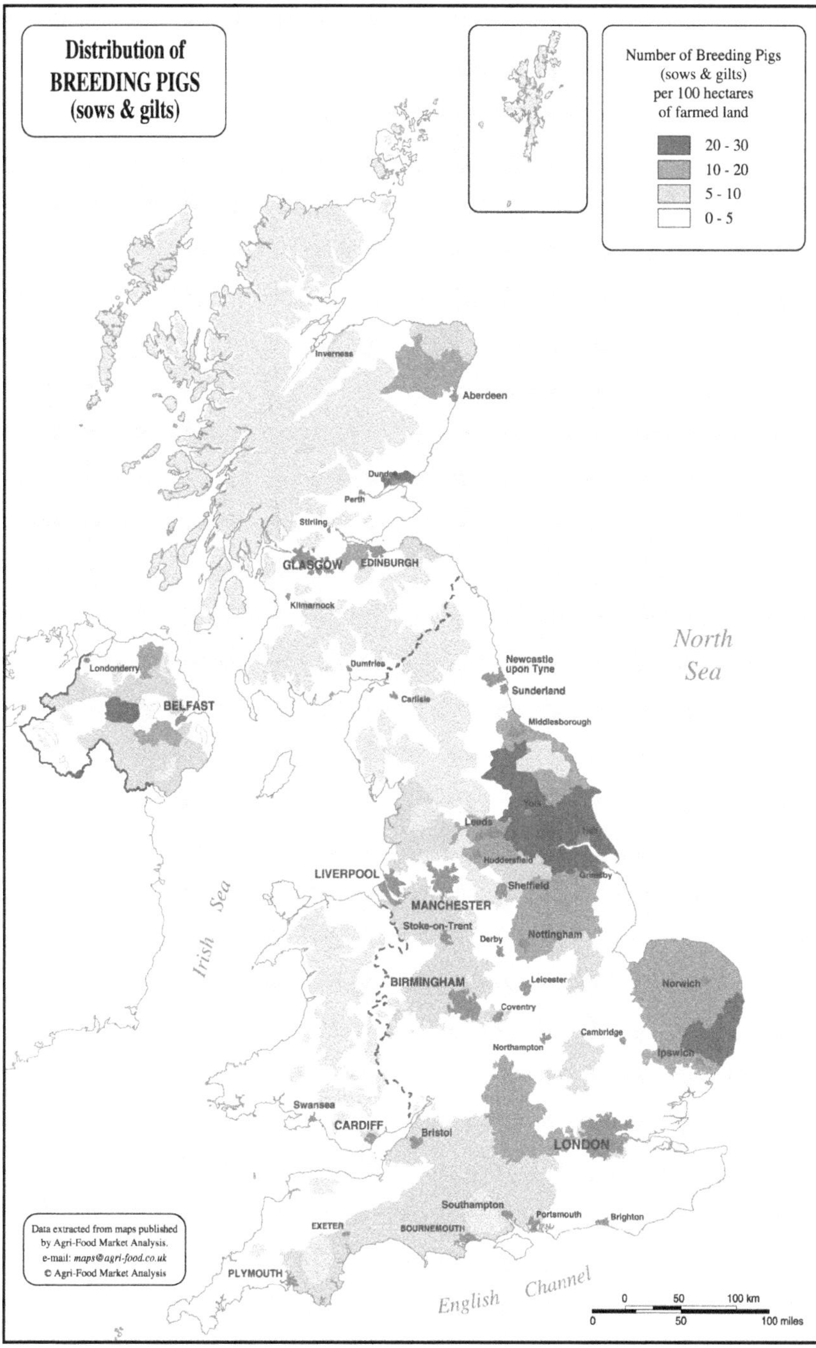
Distribution of
BREEDING PIGS
(sows & gilts)
Number of Breeding Pigs
(sows & gilts)
per 100 hectares
of farmed land
20 - 30
10 - 20
5 - 10
0 - 5
Inverness
Aberdeen
Dundee
Perth
Stirling
GLASGOW
EDINBURGH
Kilmarnock
Dumfries
Carlisle
Newcastle upon Tyne
Sunderland
Middlesborough
North Sea
Londonderry
BELFAST
York
Leeds
Hull
Huddersfield
Grimsby
LIVERPOOL
Sheffield
MANCHESTER
Stoke-on-Trent
Derby
Nottingham
Irish Sea
BIRMINGHAM
Leicester
Norwich
Coventry
Cambridge
Northampton
Ipswich
Swansea
CARDIFF
Bristol
LONDON
Southampton
Portsmouth
Brighton
EXETER
BOURNEMOUTH
PLYMOUTH
English Channel
0 50 100 km
0 50 100 miles
Data extracted from maps published
by Agri-Food Market Analysis.
e-mail: maps@agri-food.co.uk
© Agri-Food Market Analysis

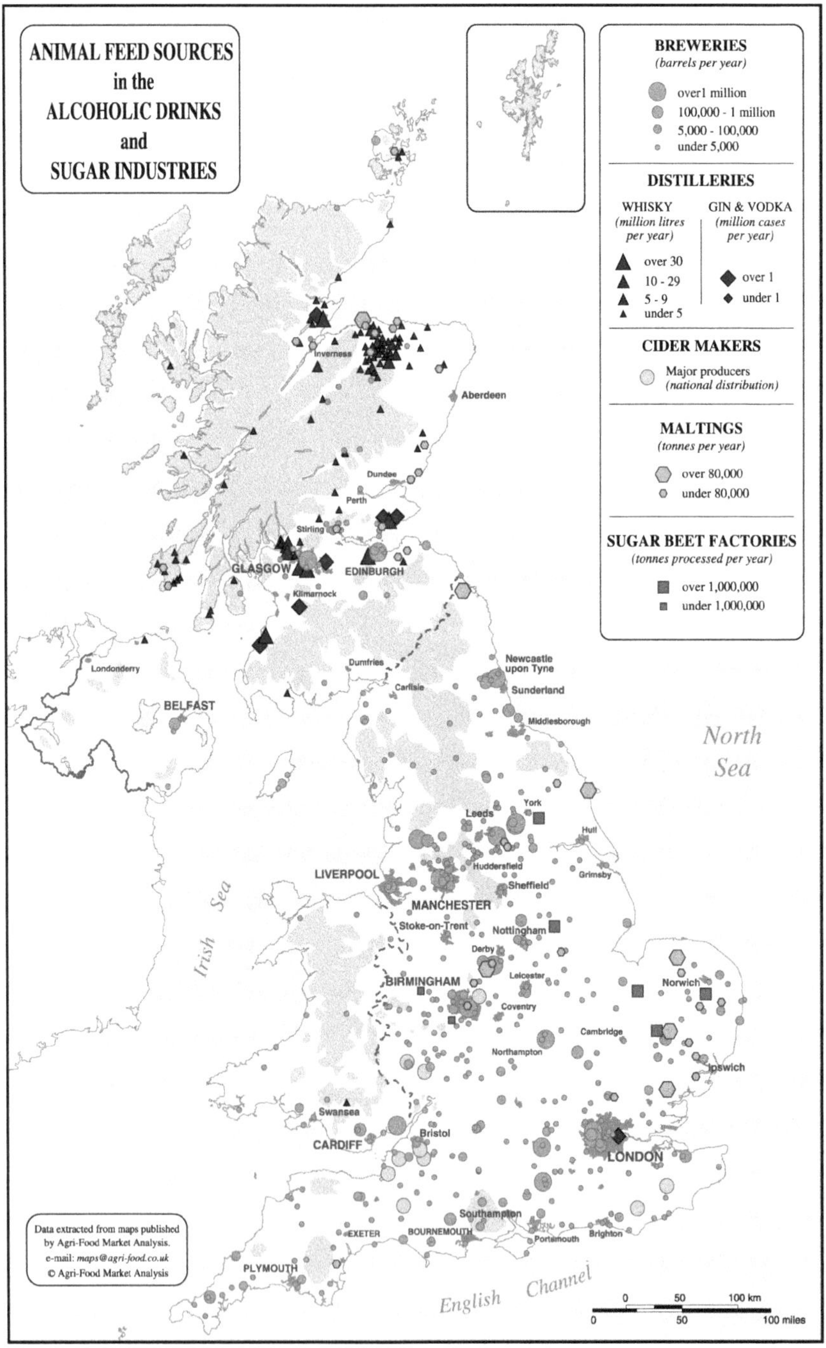

ANIMAL FEED SOURCES in the ALCOHOLIC DRINKS and SUGAR INDUSTRIES
BREWERIES
(barrels per year)
over1 million
100,000 - 1 million
5,000 - 100,000
under 5,000
DISTILLERIES
WHISKY
(million litres per year)
over 30
10 - 29
5 - 9
under 5
GIN & VODKA
(million cases per year)
over 1
under 1
CIDER MAKERS
Major producers
(national distribution)
MALTINGS
(tonnes per year)
over 80,000
under 80,000
SUGAR BEET FACTORIES
(tonnes processed per year)
over 1,000,000
under 1,000,000
Inverness
Aberdeen
Dundee
Perth
Stirling
GLASGOW
EDINBURGH
Kilmarnock
Dumfries
Carlisle
Londonderry
BELFAST
Newcastle upon Tyne
Sunderland
Middlesborough
North Sea
York
Leeds
Hull
Huddersfield
Grimsby
LIVERPOOL
Sheffield
MANCHESTER
Irish Sea
Stoke-on-Trent
Nottingham
Derby
Leicester
BIRMINGHAM
Norwich
Coventry
Cambridge
Northampton
Ipswich
Swansea
Bristol
CARDIFF
LONDON
Southampton
EXETER
BOURNEMOUTH
Portsmouth
Brighton
PLYMOUTH
English Channel
Data extracted from maps published by Agri-Food Market Analysis.
e-mail: maps@agri-food.co.uk
© Agri-Food Market Analysis
0 50 100 km
0 50 100 miles

1 Introduction

Up to two million tonnes of moist feeds are purchased by British livestock producers every year for feeding to cattle, pigs, and sheep, and a further one million tonnes of liquids are put to a similar purpose (Table 1). Still more of such feed materials are available in dry form, either separated from dry raw materials during the early stages of processing, or dried at the end of the production process. A small proportion of feed materials is dried as part of the manufacturing process. In between the extremes of high and low moisture feeds, there are syrups that have been concentrated, usually by evaporation, but not taken to full dryness. Despite significant differences in form and nutritional value, all of these feeds are essentially similar – they are derived from raw materials of the highest quality, that have been selected to meet the requirements of a discerning market. All of them represent the co-products of food and drinks manufacture. Some co-products are used for purposes other than feed, but they are outside the scope of this book, which is solely devoted to co-product feeds.

Table 1. Quantities of Co-product Feeds

Industry	***Annual Volume '000 tonnes***	
	Moist feeds	*Liquid feeds*
Apple processing	20	
Baking	125	
Brewing (yeast/beer)	750	150
Citrus & tropical fruits	42	
Distilling	343	20
Maize fractionation	***	some
Milk processing	-	517
Potato processing	153	119
Sugar beet extraction	190	-
Wheat fractionation		~200
Others	25-30	10
Est. Total	~1700	>1000

Sources: various

N.B. Figures represent known sources, and the total is probably conservative

*** From autumn 2001, it is expected that an annual total of 150,000t of moist maize gluten feed will become available

Large cattle need a lot of feed
(courtesy: James & Son (Australia) Pty Ltd)

British breweries have supplied ~700,000 tonnes of feed every year for the last hundred years
(courtesy: James & Son (Grain Merchants) Ltd)

Some co-product feeds are produced in the traditional industries of baking, brewing, cheese making and distilling. Others come from more recent food and drink processes, such as the conversion of potatoes into convenience and snack foods, and the production of fruit juice and ready-made salads. Wherever a food raw material is processed there are inevitably fractions that are not required in the finished product. Greater precision in the specification of human foods – sometimes in relation to largely cosmetic features such as size, shape and colour – results in a greater proportion of the raw material not being used. Much of this non-selected fraction becomes available as animal feed, and the co-product feed market continues to grow.

It is clearly of paramount importance for the basic raw materials to be of the highest quality and safety if they are to be used in the production of food and drink destined for the human market. All raw material suppliers are subject to stringent audits, carried out at regular intervals by independent inspectors commissioned by the major food retailers. Those audits provide certified assurance that the raw materials were produced by good farming practice, stored safely in accordance with MAFF guidelines (MAFF, 1995), and transported to the factory in vehicles that conformed to the required standards. In many cases transport would comply with the UKASTA Code of Practice for Road Haulage (UKASTA, 2000). From the receipt of raw materials at a modern food processing plant, compliance with the highest standards of food safety and hygiene is expected, but today's consumers require verification - proof that risks have been appraised by a HACCP procedure and every effort has been made to minimise their potential impact. The British Retail Consortium now has immense influence on the standards that are applied throughout the entire food chain and, within the food processing factory, quality control procedures must be in place to monitor all operations with full records of the process being retained for independent audit.

When co-product fractions leave the production line many of them fall within the compass of the BFBi Code of Practice (ABTA, 1998; BFBi, 2000), which requires moist feed materials to be moved, stored and dispatched with the care one would expect of items that remain part of the food chain. (Since the Lamming Report of 1992, it has been formally recognised that animal feeds form an integral part of the human food chain, for the purposes of food safety.) From time to time, new processes will be introduced that recover all or some of the co-product fraction, and put it to more profitable use in the production of other food products. Thus, in time, some co-product fractions may disappear, diminish in volume or change in composition, but the fundamental quality of this range of feed materials never alters – they remain feeds with an enviable provenance.

Feed definitions

The legal definition of a feed material is very broad. The Feeding Stuffs Regulations 2000 (England)[1] - citing directly from EU legislation - provides the following guidance ...

> "any product of vegetable or animal origin, in its natural state, fresh or preserved; any product derived from such a product by industrial processing; or any organic or inorganic substance (whether or not containing any additive) and for use in oral feeding to pet animals or farmed creatures".

The essence of this definition is that any material may be used as a feed, with the proviso that it does not contain an inappropriate level of any undesirable substance. An additional clause, requiring a feed material to be "sound and genuine, and of merchantable quality" would seem to provide a valuable safeguard, but these terms are themselves rather broad. In the past, the all-encompassing nature of this definition has not prevented the use of domestic waste (solid urban waste from refuse collections) in Italy or, allegedly, the inclusion of human sewage in compound feeds in France. Following the reporting of these incidents, it proved necessary to make specific exclusion of such unlikely and undesirable materials. The recent foot and mouth epidemic in the UK, and its association with the swill feeding of pigs, has again highlighted old practices which must be deemed incompatible with the objective of maintaining the highest levels of food safety throughout the food chain.

The EU Commission has established a non-exclusive list of the main feed materials, and this is re-stated in Schedule 2 Part II of The Feeding Stuffs Regulations 2000. Part III of this schedule allows for the use of other feed materials, providing that adequate identification and analytical definition is supplied. However, the Commission is currently considering the need to tighten the control of feed materials. The European Parliament has expressed its support for a "positive list", and the Commission is setting up a feasibility study of this subject. The fundamental difference between the existing non-exclusive list and the prospective positive list of feed materials is that, whilst the former allows for the use of other feeds, the latter will make the use of any unlisted feed material far more difficult. A date of December 2002 has been set, by which proposals for a positive list must be tabled, though the details will presumably depend on the conclusions of the feasibility study.

---

[1]Equivalent regulations were introduced in Scotland and Wales in 2001

In order to cope with the consequences of technological changes at processing sites, any form of exclusive list would need to include provision for additional feeds to be added, though experience suggests that any amendment would not be achieved quickly. Thus, whenever such proposals are implemented, it would seem advisable to ensure that all feed materials are included in the initial listing. The scale of the task may be gauged by comparison of the published list of 166 main feed materials with a draft list of 612 feeds, drawn up in 1990, and purporting to represent the ingredients used in compound feeds in the Community (EC, 1990). This latter document, which aimed to exclude any feed not used in compound feeds, was by definition an understatement of the total number of feed materials in use on EU farms. Scrutiny of the list reveals that, with the exception of "vegetables and potatoes", two seemingly atypical entries added to the bottom of the list, all of the listed feeds are of materials presented in dry form. Thus, if the proposed positive list is not to exclude many feeds already in use on farms at the present time, it will be essential to ensure that moist and liquid feed materials are included alongside a considerably extended list of dry feed materials.

Products and by-products

Feed materials are commonly classified as either products or by-products, but in terms of animal feeding the distinction is of no consequence. The classification is not universally adopted and the sub-division appears to be inconsistent. Thus in The Feeding Stuffs Regulations 2000, whole cereal grains and oilseeds remain outside the classification, but so do grape pips, cereal straws and cocoa husks. Flaked cereals and a number of fish, meat and powdered milk fractions are identified as products, as are seaweed, bone and hydrolysed feather meals, together with ground fodder rice and alkali-treated straw. Greaves is referred to as a "residual product", which appears indecisive, whilst by-products include materials that have been subjected to a range of both physical and chemical processes. Examples include whole grains that have simply been fractured such as broken rice, fractions that have been separated, like wheat germ, and extracted materials such as the pulps of citrus fruit, tomatoes and sugar beet. Grape pulp, which is subjected to an even greater degree of processing - including fermentation as well as extraction and some fractionation - is classified as neither a product nor a by-product.

If logic can be applied to the classification, it would appear that the distinction between products and by-products usually takes place outside the animal feed industry, and it typically reflects the objectives of other parties. Thus millers

---

[II] At the time of writing, in June 2001, the foot and mouth epidemic still continued

provide flour for the baker, and the "by-products of the milling process" - wheat middlings and bran - are supplied as feeds for use by cattle, sheep and pigs. The term "flour miller" would appear to brook no argument with regard to the relative status of the end-products – flour production is clearly the miller's principal objective – but the animal feed industry has no direct interest in flour whatsoever. In a parallel process, brewers remove much of the starch and sugars from barley and malt for use as substrates in their fermentation process, and this leaves a solid residue that has always been referred to as "spent grains". The term suggests that this is, self-evidently, a by-product material and it is subsequently used as a feed for ruminant livestock. But like that of the miller, the role of the the brewer in animal feeding is limited to that of feed supplier, and since the animal feed industry rarely makes use of the products of milling and brewing - namely flour and beer - the classification is of little relevance. Arguably, it would seem inappropriate to classify feeds in terms of their utility for other, non-feeding, purposes and wheat middlings, bran and brewers' grains could more realistically be referred to as feed materials from the flour milling and brewing industries. This would be suitably specific, and preferable to the use of hand-me-down terms that may seem to imply that they are second grade materials from another industry.

In nutritional terms, by-product feeds are demonstrably not "second grade" materials. In fact the product/by-product classification provides a very inaccurate guide to feeding value. The removal of starch in brewing leaves a material that is richer in other valuable components. On a dry matter basis, brewers' (spent) grains have a protein level that is typically more than twice as high as that of the original cereal grain and a gross energy value that is some 12-18 per cent higher (MAFF, 1990). Admittedly, the fibre content of the feed is higher and the digestibility of the energy is somewhat reduced, but the ruminant animal – to which this material will largely be fed - has evolved a digestive system that makes efficient use of fibrous feeds. The advance of enzyme technology also opens up the prospect that the relatively low digestibility of such feeds may be improved. Thus the quality of the feed may be represented as merely a temporary challenge to the livestock industry, and other processing methods may eventually be found that will impart a higher feeding value. A further example of the pointlessness of the classification is afforded by the extraction of sugar beet, where the removal of the sugar leaves a palatable pulp of similar digestible energy value to the original root, on a dry weight basis. Contrastingly, the energy value of the pulp is largely in the form of digestible fibre whereas sucrose is the energy store of the unextracted beet, but for ruminant feeding that may place the pulp at an advantage. The pulp provides a similar amount of energy to the sugar beet root but in a form that is less likely to cause acidosis – a serious consequence that may result from the rapid digestion of sucrose. The by-product pulp is consequently a safer feed for cattle and sheep.

There are major variations in the nutritional value of different plant species and these also serve to emphasise the futility for animal feed purposes of the distinction between products and by-products. Within the cereals, for instance, most maize by-products are of higher digestible energy value than unclassified oats and barley grain, and of substantially higher value than the products fodder rice and treated straw. A similar discrepancy can be seen with other types of feed, and an obvious example would be the range of high protein materials that result from the extraction of oil from oilseeds. All of the resulting by-product cakes and meals are much richer in protein than the original oilseed and, in many cases, they are richer in protein than specifically grown protein products such as peas, beans and sunflower seed.

Co-products and wastes

The terms "co-product" and "co-product feed" were adopted by the relevant parts of the animal feed industry in the 1990's in response to legislative proposals that sought to classify anything other than the primary products of food processing as "wastes". It is clear from the previous discussion of products and by-products that many feed materials would not be adjudged to be the primary products of the processing facility where they are produced, and almost half the entries in the EC feed materials list are described as by-products.

The definition of waste (OJ No L 78/32) is even broader than the definition of feedingstuff. Sixteen categories of waste were identified in the Directive published in March 1991, with the final category being a catch-all to include "any materials, substances or products which are not contained in the above categories". Even without that final category, the overlap between this list and the list of straight feedingstuffs – the forerunner of the feed materials list - was substantial. And the definitive ruling that a material constituted either a feed or a waste could only be determined, possibly after prolonged delay, by the European Court of Justice. The uncertainty brought by this Directive was the cause of great concern to feed merchants, and to the manufacturing facilities whose products now seemed to be under threat. Their feeds had been supplied to a large and mostly satisfied group of customers over many generations, and now their status and even their continued marketability had been put in doubt.

The implications of the original waste proposals were that any farmer wishing to continue to use such feed materials would need to obtain a waste management licence, and any merchant or haulier would need a certificate of registration as a waste carrier. Every delivery would need to be accompanied by a waste transfer note that identified the source, the carrier and the nature of the waste consignment. To the food and drinks industries, engaged in a partnership with merchants to supply feeds to farms, there was a stigma attached to the relegation of this activity from feed supply to waste handling. But of potentially greater commercial significance, was the likely impedence, if not complete inhibition, of the marketing of these materials to a wide customer base. The regulations insisted that "controlled wastes" could be delivered only to a licensed site, or to a site for which an exemption had been registered, and the Local Waste Regulation Authority required pre-notification.

Representations on behalf of a number of feed suppliers were made to Ministers at the Department of the Environment, and to their equivalent in The Scottish Office. Satisfactory assurances were provided by three Ministers in 1993 ... "no materials from whisky distilling operations (Scottish Office, DOE 1), or industries such as brewing (DOE 2), which are not at present wastes will be reclassified as wastes as a result of our proposed Regulations on waste management licensing". The crucial decision, of whether a material was a waste or not, was considered to depend on the attitude of the producer; whether he could be regarded as having "discarded, intended to discard or been required to discard the material in question". This return of the ball to the producers' court was most welcome, even if it did require some factory managers to be more careful with their terminology. Even today, the term "waste" occasionally slips out in unguarded conversation, despite major efforts by factories to upgrade their storage and handling facilities, and the extension of quality control procedures to cover this part of the operations.

Although the term "co-product" was introduced in order to differentiate useful, safe and wholesome feed materials from wastes, its use does recognise a fundamental truth. In any food processing facility, the fraction that is destined to become a feed is of equal status to the primary product in terms of both origin and safety, and its nutritional value may be even be higher. In other important respects, the co-product is unquestionably different – it is not the prime objective of the processor and it does not have equal financial value. But if food safety and traceability are the principal focus of interest, co-products have an identical background to any other fraction separated from the same raw material.

Co-product feeds originate in British food and drink factories, and thus their provenance cannot be surpassed as feed materials fit for recycling into the food chain. The careful selection and control of the raw materials that make them suitable for use by food and drink manufacturers, automatically render the associated co-products of their processing operations equally suitable. Thus consumers need have no concerns about the safety and suitability of such feeds at the point of production. Co-product feeds offer advantages to livestock consumers too, in terms of the attractiveness of the feed supply. Processors often have limited facilities for the storage of co-products, and their continual removal from the premises is important if factory production is not to be interrupted. This potential problem for the factory becomes an asset for the livestock farmer, since the regular removal of co-products results in the continued availability of supplies of fresh feed.

Codes of Practice and Feed Assurance

Feed legislation continues to grow in both volume and complexity, though popular perception appears to be that it provides an inadequate degree of control. Over the last ten to fifteen years, there have been a series of major food scares – several of which have been associated with feed – and these have rocked public confidence. The need for additional reassurance was acknowledged by UKASTA Chief Executive, Jim Reed (UKASTA, 2001), who noted that "while improved regulations could provide feed and food safety, the main lesson from the BSE crisis was that the law alone was not enough to restore consumer confidence in UK-produced livestock products". Over the last four years UKASTA, the principal organisation representing feed compounders and the vast bulk of the animal feed trade in the UK, has been developing a number of Codes of Practice. The essence of these codes is a definition of best practices that ensure that the high quality of feed materials is safeguarded from collection point to safe delivery area on farm. Any part of the feed industry wishing to claim compliance with the Codes must adopt these practices in full. Since their introduction, the UKASTA codes, listed below, have contributed significantly to a lifting of standards within various sectors of the supply industry:-

- UKASTA Code of Practice for Road Haulage (of combinable crops, animal feed materials and as-grown seeds).
- UKASTA Code of Practice for the manufacture of safe compound animal feedingstuffs – and an abbreviated version for use by merchants and distributors who do not manufacture compound feeds.
- UKASTA Code of Practice for the supply, storage and packaging of animal feed materials which are destined for farm use.

The moist feed industry, represented by the merchants who formed the brewers' grains committee of the Allied Brewery Traders Association, recognised that the UKASTA codes did not adequately cover the moist feeds that were the bulk of their business, and that in certain aspects these codes were inappropriate for moist feeds. However, there were negative implications in being outside the UKASTA codes that were being applied to most other parts of the feed industry. To third parties, exclusion may appear to be an indication that these feeds were unapproved. Thus in 1999, the ABTA (now BFBi) launched its own Code of Practice for the Supply of Moist Feeds.

Ahead of all these efforts to improve food safety was the Freedom Food's initiative launched by the RSPCA in 1994 to improve the welfare of farm animals and to provide consumers with the opportunity to purchase clearly identified, welfare-friendly products. Standards were set for the rearing and handling of farm animals throughout their lives and accredited farms were permitted to use the Freedom Food trademark. Inspection then proceeded along the food supply chain through processors, manufacturers and packers so that consumers could be assured that trademarked products on supermarket shelves could be traced back to accredited producers.

Although Freedom Food had only a minor and indirect effect on the animal feed industry, the initiative opened the way for other outside influences to be brought to bear. Pressure began to be applied from the consumer end of the feed chain, where the retailers and food processors began to take an increasing interest in how their livestock products had been produced, and to dictate which feeds and feeding practices were acceptable and which were not. Several individual dairy companies produced their own Codes of Practice, with the Unigate Superior Stockmanship Scheme (1997) being a comprehensive guide that others followed to a variable extent. Meat producers were also encouraged by supermarket buyers to join an assurance scheme (Austin 1996), and Farm Assured British Pigs (FAB Pigs) became the first of its kind. This was followed by a similar scheme to be operated in Scotland, known as the Scottish Pig Industry Initiative (SPII). Other red meat producers were encouraged to join Farm Assured British Beef and Lamb (FABBL), which aimed to represent all aspects of the British beef and lamb industries. There were alternatives to this organisation too, in the form of the Scottish Quality Beef and Lamb Association (SQBLA) and the Welsh Quality Lamb (WQL) scheme, both of which set out to establish standards, which promoted meat quality alongside national identity.

In addition to these initiatives by the feed and livestock industries, supermarkets also launched their own schemes. Sainsbury's, Tesco, Marks & Spencer and others promoted partnership schemes with meat producers or with an approved abattoir group (Wright, 1996; Clarke, 1996). The level of control that supermarkets sought to exercise differed between the companies, but the strongest link was that proposed by the largest food retailer, Tesco, whose ultimate objective was reported to be meat from its own producer groups, produced by animals fed on Tesco-branded feed (Clarke, 1996). Whether through such individual initiatives or collectively, the supermarkets seem certain to retain a major influence on future trends in animal production and livestock nutrition. Consumers apparently perceive supermarkets as sympathetic to their desire for greater quality, safety and integrity in food production, and they identify supermarkets as the enforcers of food safety (Ratcliff, 2000). To date, the public appears not to have appreciated the huge advances that have been made by the feed and livestock industries in recent years. When foot and mouth disease revisited Britain in 2001 after a twenty-year gap, there was much uninformed comment on the use of "food wastes" and uncontrolled farming practices. Once again the media focused on the bad rather than the good, leaving farmers with the task of rebuilding consumer confidence after an epidemic that devastated many individuals and great tracts of the country, and which was not of the farmers' making.

Consolidation: Codes of Practice were widely considered by all parts of the feed and livestock industries to be a positive response to consumer concerns about food safety, but their proliferation added an element of confusion. There was a profusion of codes differing in both detail and geographical coverage, and whilst some were weighted, like the Freedom Food scheme, towards animal welfare others were aimed solely at food safety. An element of competition between the codes was also evident - an essential aspect for the supermarkets - but in national terms this did little to satisfy public concern about the wider issues of food quality and safety. Echoing this sentiment, Lord Lindsay, Chairman of Assured British Meat noted that "the integrity of a (safety assurance) scheme must be immune to commercial and competitive pressures" (ABM, 1998). Eventually, the need for co-operation between the proponents of the many schemes was recognised as beneficial – together the various parties could establish a reputation for British food that would outweigh any small advantages that may be attained separately.

At farm level, the response to the codes established by retailers, processors and the feed industry has been the development of the National Dairy Farm Assurance Scheme (NDFAS, 1998a and b). This scheme has incorporated the best elements of other codes and adapted them to fit farm practice. NDFAS aims to cover both the issues of animal welfare and food safety, and it has gained wide support from the British Cattle Veterinary Association, the Dairy Industry Federation, the National Farmers Union and the Federation of Milk Groups. Most milk processors now appear to have adopted the NDFAS in place of their own codes and consumers are reassured that all British milk producers are working to the same, stringent standards. With regard to hygiene and food safety, the stated aim of the NDFAS is "to provide reassurance for customers in terms of the safety and high quality of the milk produced". In relation to the feed supply, the NDFAS requires that: "all feed and water provided for dairy cattle must be of an appropriate quality for a properly balanced diet". Supporting those broad objectives there is a lengthy list of well-defined details that require compliance by both the farmer and his feed suppliers.

The meat industry has set out to accomplish an even wider co-ordination of previously disparate parties in a scheme that single-mindedly targets food safety. Feed producers, farmers, transporters, auctioneers, abattoirs, processors, wholesalers, butchers and caterers have all been persuaded of the merits of linking their efforts towards a common objective. Assured British Meat (ABM), founded with Government backing as an independent, non-profit-making body in 1997, quickly established a predominance within the red meat industry that gave credence to the ABM claim that it would "assure food safety throughout every part of the food chain" (ABM, 1998). Notably, ABM's total commitment is to food safety, and animal welfare and environmental issues are included only when they have a bearing on food safety, or on the wholesomeness and acceptability of a product. A basic tenet of ABM assurance is that it is driven by best practice and public expectation rather than the legal minimum, and it is significant that the standards set by this scheme are supported by both retail and consumer organisations.

Continuous assurance throughout the meat supply chain

(courtesy: EFSIS)

Even before consolidation, it became quickly apparent that Codes of Practice were of little value unless checks were made to ensure that the appropriate standards were being met, and that sanctions would need to be imposed on those failing to reach or to sustain the required level. The need for critical inspection and appraisal was accepted and independent auditing was established. However, with annual accreditation, it may have been necessary to undertake multiple audits of the same production site in order to satisfy a number of differing requirements. Consolidation has brought us close to the point where one audit can be used to satisfy the needs of a wide range of interested parties.

Thus UKASTA has launched its feed assurance scheme (UFAS), and both ABM and NDFAS have accepted that this scheme provides an appropriate standard for the feeds used by their suppliers. The BFBi Code of Practice has been recognised by UKASTA as the definitive standard for moist feeds, and BFBi members are aiming to achieve UFAS accreditation. Compliance with the requirements of UFAS will be audited by the European Food Safety and Inspection Service (EFSIS Certification), and accreditation will be awarded to a recognised European quality standard, EN45011. The certification system itself will be assessed by UKAS, the United Kingdom Accreditation Service and, through all of these links, the consumer can be given assurance that the safety of the food chain is completely and continuously audited by trained and independent specialists from start to finish.

Certification Mark

The link between food safety and animal welfare is also strengthening. Freedom Food recently reported "great progress" towards the achievement of full farm assurance (Freedom Food, 2000), with the addition of elements of food safety and hygiene to its core of welfare requirements. Two-way training is being carried out to enable assessors from Freedom Food to carry out inspections on behalf of SPII, SQBLA, and NDFAS, and vice versa. Discussions about similar arrangements are also continuing with FABBL and ABP and it seems probable that, in the foreseeable future, assurance of the broad objectives of each scheme will be secured across a wide section of the British livestock industry.

Much has already been achieved, and the consumer can begin to re-establish the confidence that may have been lost in food from the British livestock industry. He or she can feel safe in the knowledge that appropriate standards are being applied, by a high proportion of those involved, at every stage of the business, and that the verification of food safety and quality is being obtained at every step.

Greener pastures: There is a gap in the feed assurance coverage that stands out oddly when compared with the comprehensive control of all other aspects of feed supply. Whilst feed materials, roots, forage replacers, and both home-mixed and proprietary compound feeds are all regulated, recorded and warrantied, forages are not yet covered by any Code of Practice. For ruminant animals, this represents a substantial area of concern; a dairy cow may consume half of its dry matter intake in the form of forage, whilst extensively-reared beef cattle may rely on forage to a greater extent. Within a short space of time, the assurance that the feed industry offers to British livestock producers will be the envy of the world, but forages produced on farm, and sometimes moved between farms, have some way to go to catch up. Grass may be green, natural and even organic, but there appears to be a need to introduce codes relating to the growth, harvest and feeding of pasture and conserved crops in order to demonstrate due care and to limit contamination of the crop.

Logistics

The location of food processing facilities may have been selected for a number of reasons, and the commercial justification for such decisions would usually be sound. Location may owe something to the historical development of the industry, to the availability of suitable quality raw materials and personnel, or to the proximity of the principal markets. For example, breweries were originally built where there was a good quality water supply and a ready demand for the product, potato processors were located in areas where potatoes were grown and, most famously, malt whisky producers believed that their product was so imbued with qualities unique to the area that relocation was almost impossible by definition. Significantly, location was never dictated by the availability of appropriate outlets for the co-products and, as production volumes have grown, this has led to increasing difficulty and burgeoning transport costs.

Haulage costs are a significant item for all industries, and the burden continues to grow. Despite costs that were claimed to be the highest in Europe, the declared policy of the British Government, until 2001, was to impose continual increases in fuel tax as a form of environmental penalty on the use of fossil fuel – "the fuel escalator mechanism". Although this policy has apparently been put on hold, transport costs remain high. All carriers of animal feed share the burden equally, but the cost impacts most significantly on those carrying high moisture feeds – in reality, they haul large and, for the farmer, largely unnecessary volumes of water around the country. Whilst cereal grain and compound feeds comprise 86-88 per cent dry matter, many moist co-products have a dry matter content of 25 per cent or less, and some liquid feeds destined for the pig market have as little as 5 per cent solids. Thus although the transport charge on a typical 24 tonne load of grain can be spread over 20 tonnes of dry matter, the same charge has to be carried by a mere 5.5 tonnes of dry matter in brewers' grains or a meagre 1.2 tonnes of whey dry matter.

The maps printed at the front of this book show the concentration of dairy herds, beef cows and pigs in the UK, and the location of the production sites of the sugar beet, brewing, distilling and cider-making industries. Whilst there remains a strong geographical link between brewing and dairying in the West Midlands, this is exceptional. The concentration of brewing in the South-East matches the density of the human population of this area, but the market for brewery co-products is in the South West, some 150-200 miles away. A similar conclusion could also be drawn with regard to the location of the principal sources of potato co-products, which are mainly to be found on the eastern side of the country. For liquid-fed pigs, the largest source of supply is at Greenwich in the heart of London – not only is this distant but the supply is constantly impeded by congested roads from the pig concentrations of Norfolk, Yorkshire and the South-West. Most distant of all are the distilleries of Central and North-East Scotland – not to mention those of the Western Isles - which lie hundreds of miles from the principal livestock areas where their co-product feeds can be put to effective use.

Liquid feeds bear an additional burden compared with those supplied in solid form, because of their need to be carried in road tankers as opposed to bulk trailers. Since tanker transport is significantly more expensive, this adds appreciably to the delivery cost. Where the liquid is a feed of low dry matter content – as in the case of some of the milk co-products – the additional cost bears down more heavily on the low nutrient value of each load. There is an additional factor that puts moist and particularly liquid feeds at a further disadvantage. Regular vehicle cleaning is an essential part of a warrantied feed supply, but moist feeds – often with a higher oil content than dry alternatives – represent a more difficult and inevitably more expensive cleaning challenge. Tanker cleaning is especially difficult, and it predictably adds more expense to the task of upholding the necessary hygiene standards.

A reduction in the water content of feed would be of major benefit to moist and liquid feed merchants, and may have desirable implications for farmers in respect of storage costs. However, their enthusiam is not necessarily shared by the moist feed producer, since dewatering may result in the release of a nutrient-rich juice that becomes an environmental burden for the factory. In olden times, such liquids could be disposed of as effluents; viz: put down the drain, discharged into rivers or piped out to sea. Although such practices still continue, they are governed by "consent levels" agreed with the water authorities and they come at a cost. Drying may also be an expensive option, since the removal of water often results in little change in the monetary value of the feed dry matter. This implies that the drying cost has to be justified by reductions in the costs of haulage, storage and effluent treatment. For distant production sites, such as those in the North of Scotland, drying may represent the most economic choice.

Dewatering is rarely as easy as it may sound; for instance the co-product of steam-peeling in the potato factory contains about 88 per cent water, but much of that water is chemically bound and is very difficult to remove except by drying. Liquid co-products can be concentrated by evaporation, but, unless they are to be dried completely, a progressive increase in viscosity puts a limit on the extent to which this process can be continued. Anyone who has marketed such syrups to farms knows of at least one instance where a hot fluid was delivered into a farm tank that was almost impossible to remove once cold. Special pipes and pumps may be used to aid the flow and enzymes - particularly ß-glucanases - have also been employed to reduce viscosity, though with limited commercial success.

Storage

Dry feeds need to be stored in a bin or in an indoor bunker, where the material can be kept dry and free from vermin or domestic animals. Codes of Practice now insist on the use of an impermeable floor and the use of retaining walls to prevent a feed material becoming contaminated. Stock rotation and cleaning routines are also encouraged to prevent the build-up of dust and stale material (NDFAS, 1998b).

Moist feeds are usually perishable and need to be ensiled and sealed effectively to reduce the extent of aerobic decay. Short-term storage is often carried out less carefully, but this is likely to result in both visible and invisible loss of digestible nutrients. Sheeting may protect the stored material from rain and the attentions of birds and mammals but, unless the sheet is held in contact with the surface of the underlying feed, it will merely provide a humid atmosphere in which aerobic micro-organisms are likely to flourish. Many feeds stored in this way are likely to lose effluent and to constitute a potential environmental threat unless collected and disposed of safely. However, the development of co-storage practices involving dry, absorptive materials such as citrus or sugar beet pulp has been of substantial benefit.

Liquid feeds need to be stored in dedicated tanks and transferred to pipeline systems by an appropriate pump. Viscous liquids may require a positive displacement pump to effect satisfactory transfer, though heavily viscous fluids such as molasses are sometimes fed by gravity into mixing wagons stationed beneath. Galvanised tanks are not suitable for the storage of acid co-products, such as whey, potato feed or Greenwich Gold, because the zinc coating dissolves and the zinc content of the feed may reach toxic levels within a few days (ADAS, 1988). Concrete tanks need to be painted to protect them from corrosion. Tank identification is important to indicate the location of each co-product, and a device to indicate the spare capacity in a tank helps to avoid oversupply and consequent spillage. A means of mixing the co-product, in-silo, would be of considerable value for some liquid co-products, such as brewers' yeast, that have a propensity to separate into layers. Stirring the ingredients prior to use would allow the ration to be mixed as formulated, and reduce the risk of a solidifying fraction blocking the pipes. A vent in the top of the tank would provide a safety valve to allow the release of gases produced as a result of fermentation. Bunds or drainage channels should be constructed in order to contain spillage.

Scope of this book

The scope of this book encompasses all of the co-product feeds that become available when high moisture raw materials are processed, and dry raw materials are subjected to a form of wet processing. A total of 81 feeds are described, many of which have not been defined previously in animal feed literature. Some of these feeds are dried at, or near to, the production site, and these feeds tend to be relatively well known. Others are marketed in a moist form and, although some of them are part of a long tradition, none of these feeds is included in the EC list of the main feed materials. Yet their combined volume is substantial by any standards. An estimate of the volume of co-product feeds marketed in a high moisture form is given in the tabulated section of this book.

Proprietary names have been used in order to avoid ambiguity, and a full description of their derivation is provided. It is expected that the precise definition of such feeds, in both words and figures, will encourage greater confidence in their use on farm. Hitherto, co-product feeds have tended to be ill-defined and often regarded as "variable". Whilst accepting that all feed materials have a degree of variability in both composition and nutritive value, it seems reasonable to expect that the better definition of co-product feeds will reduce their reputed variability. One example will serve to illustrate the point: potato processing waste is listed in UK Tables (MAFF, 1990) and, putting aside any discussion of "wastes", this term could be applied to materials ranging from potato skins (ME value 5.5 MJ/kg DM) to potato crisps (ME 22.8 MJ/kg DM). This book identifies 19 separate potato co-products that are being marketed as animal feeds, and it is hoped that the asssociated information will allow livestock farmers to make better use of them.

Co-products of animal processing: It is recognised that there are other sources of co-product feeds – for instance, the author has previously reviewed the nutritional merits of a number of animal processing co-products - feathermeal, bloodmeal, animal fat and meat and bonemeal (Crawshaw, 1992b, 1994, 1994b, 1995). However, since the development of the BSE disease in cattle, and the acceptance that the major route of transmission was through ruminant-derived meat and bonemeal (MBM), the use of animal co-products as feedingstuffs has been increasingly curtailed. What began as a ban on the feeding of ruminant protein to ruminant animals has been extended to a prohibition of the use of any mammalian MBM to any farmed livestock.

The Food Standards Agency review of BSE controls in December 2000 (FSA, 2000), expressed the opinion that there was no scope for relaxing the ban in the foreseeable future. Indeed it went further, by recommending the consideration of a complete ban on the recycling of animal material within any species, since there was a risk that this practice "could amplify a new TSE (transmissible spongiform encephalopathy) in a species". Although such a ban would still leave the theoretical possibility of using pig MBM in poultry feed, and poultry MBM in pig feed, the FSA opposed the practice. It took the view that any infectivity in pig MBM would not be inactivated in the chicken intestine, and this may in course lead to its reincorporation into pig feed – which would amount to undesirable intra-species recycling. Nor was the FSA confident that, given the complexity of the animal feed manufacturing chain and previous problems of cross-contamination, the separation of feed streams for pigs and cattle could be assured or enforced in practice. Adding a final non-scientific comment to what it clearly considered to be an unacceptably risky practice, the FSA suggested that - even if adequate controls could be developed and implemented by the rendering and animal feed industries - it was unlikely that consumers would wish to buy meat from animals fed in this way.

Thus, the use by the British livestock industry of feeds derived from animal processing is improbable at the present time, and likely to be controversial even in the longer term. That notoriety may be unfortunate, but it distinguishes such feed materials from those identified in this book. The essence of the selected co-product feeds is that they are feeds of enviable provenance. They have been plucked from food processing operations that supply food items direct to the human food market and they present no innate hazard either to the animals or to the ultimate consumers of animal products. Once they have left the direct food supply, these co-products are protected from contamination throughout the journey that carries them from the factory to the feed trough, where they re-enter the human food chain once more.

Pharmaceutical co-products: In addition to the co-products of the food and drinks industries, there is also potential feed material available from the pharmaceutical industry. Microbial biomass from industrial fermentations is typically a high protein material with a rich content of essential amino acids, and obvious potential as a feed material. But that produced as a co-product of pharmaceuticals may carry additional risks. Over the last few years, increasing concern has been expressed about the development of antibiotic resistance, and the use of antibiotics in feedingstuffs has been severely cut back. Whilst this has been occurring, it would seem retrograde and contradictory to encourage the use of any feed material that may have antibiotic properties. Thus prior to the approval of any biomass that arises as a result of antibiotic production, a detailed dossier of evidence must be presented of the product's safety.

Guidelines for the assessment of "certain products", as they are officially termed within the EU, were provided in an EC Directive (EEC, 1983), and they are extensive. 'As a general rule', the Guidelines indicate, 'all the information necessary to establish the identity of the micro-organism and the composition of the culture medium, and also the manufacturing process, characteristics, presentation, conditions of use, methods of determination and nutritional properties of the product must be provided'. A similar amount of detailed information is required in order to assess the tolerance of the product by the target species, and the risks for man and the environment which could result directly or indirectly from the use of the product. The toxicological studies required for this purpose depend on the nature of the product, the animal species concerned and the metabolism of the product in laboratory animals. It is made clear that all the studies outlined in this document may be required and, if necessary, additional information may be requested.

So onerous is the proof that is required, with no guarantee that the approval of material would not be held up by demands for further tests, that pharmaceutical companies are discouraged from starting out along this road. That may be a pragmatic assessment of the future of such materials as feed materials. In recent discussions, the author formed the opinion that it was unlikely that any material of antibiotic origin would be approved in the current climate of opinion. Thus the potential feeding value of pharmaceutical co-products seems unlikely to be realised and they are not included in this book.

2 Apple Processing

Both wild and cultivated forms of apple are believed to have grown in prehistoric times across the vast area that stretches from the Caspian Sea to the Atlantic Ocean; - the fruit was certainly known as early as 100 B.C. (Hulme and Rhodes, 1971). Apples have proved to be an enduringly popular fruit and they are now grown in many parts of the world, in both the northern and southern hemispheres. World production volumes continue to rise and the United Nations Food and Agriculture Organisation's records show a three-fold increase during the last forty years (FAO, 2000). However, while the world crop continues to expand – much of it on rootstocks developed at the East Malling Research Station in England - UK apple production has been in decline, since its peak at 650,000 tonnes in 1964. The current UK production level is only one third of that peak amount – see Table 1.

Table 1: Apple Production (mt per annum)

Year	*World*	*UK Production*
1961-63	19.9	0.5
1967-69	25.4	0.4
1977-79	33.2	0.3
1987-89	41.2	0.3
1997-99	58.2	0.2

Volumes represent 3-year averages
Source: F.A.O.

The processing of apples is a traditional practice in Britain, and particularly their conversion into the fermented drink, cider. This industry has continued to grow even while total production volumes have shrunk, with the result that the proportion of the UK apple crop that is processed for cider production has now risen to 40 per cent (Loraine Bodington, H P Bulmer Ltd; personal communication). Fresh apple juice is also being consumed in increasing quantities in Britain. The Soft Drinks Association (BSDA, personal communication) estimates that apple juice comprises some ten per cent of fruit juice consumption, and MAFF (1988) statistics record a 44 per cent growth in the fruit juice market over the last ten years. However, only a small fraction of UK apple juice is derived from apples pressed in this country (the BSDA estimate is 1.5 per cent), and thus the volume of co-products from apple juice production is only small – perhaps only 600 tonnes per annum.

Although English cider has been known since Roman times, the Normans who arrived in the eleventh century are credited with the introduction of the organised cultivation of cider orchards (NACM, 2000). The climate and soil conditions of South-West England, like those of North-West France, may have been marginal for the production of grapes, but both regions were considered ideal for the production of cider apples. A thousand years later, the UK remains both the largest producer and consumer of cider, ahead of South Africa and France, and the principal cider orchards are still largely confined to a belt that stretches across South-West England from Hereford and Worcester to Devon. Additional crops are now grown in East Anglia, the South-East and more recently in Wales. The popularity of this traditional drink has burgeoned since the 1980's, and this has led to unprecedented growth in the industry. Three thousand hectares of new cider orchards have been planted, doubling the total area, and cider production has increased from 3.4 to 5 million hectolitres per annum over the last decade. This volume will increase further when all of the new orchards reach full production. Cider orchards tend to suffer from biennialism, with substantial variation in annual production, but the volume processed by the cider industry is currently estimated to be 100-150,000 tonnes per year.

Cider is a broadly defined beverage that includes significantly different products in various parts of the world, from an unfermented apple juice in the United States to a sweet apple wine in France. The British definition allows for some variation in the ingredient mix and a significant range of product types: "a beverage obtained by the partial or complete fermentation of the juice of apples....or concentrated apple juice....with or without the addition, before or after fermentation, of sugar or potable water" (NACM, 1992). From the animal feed viewpoint, variations in the fermentation of the expressed apple juice are of no consequence for the quality of the apple pomace that is ejected from the presses. But the use of sugar and concentrated apple juice implies a proportionate decrease in the use of home-grown apples, and a reduction in the associated availability of co-products for use as feed. The typical yield of pomace is in the range of 18-24 per cent by weight of the processed fruit, and thus the annual volume is likely to be of the order of 20-35,000 tonnes. A proportion of the fresh pomace is dried for pectin extraction and is consequently diverted away from the animal feed market. In former times, much of this material would have reappeared on the market in the form of pectin-extracted fruit, but the sole pectin extraction facility in Britain has now closed and any pomace destined for this use must be exported, possibly to Denmark, Germany or France. There is consequently no pectin-extracted fruit available to the British livestock producer.

Apples are generally categorised into three types: "dessert", "cooking" or "cider" – with the first two referred to collectively as "culinary fruit". There are hundreds of varieties of each type and the rustic names of some cider varieties – such as Brown Snout, Chisel Jersey and Slack-ma-Girdle – may suggest that cider making is a quaint, old-fashioned practice. But this belies the industrial scale and technological basis on which the modern industry operates; 90 per cent of English cider is now produced by three large manufacturers. True cider apple cultivars have been selected for a high level of fermentable sugar, a relatively high concentration of tannins (which provide the characteristic flavour, or "overall mouth feel" in the industry jargon), and a fibrous structure that facilitates the expression of the apple juice. The ability of the fruit to mature during storage, converting starch to sugar without losing its crucial texture is also a desirable feature. However, cider production is a seasonal industry stretching from September to mid-December, and apples are mainly used fresh from the orchard with a minimum of storage.

Cider apples are divided into four categories according to their relative proportions of acid and tannin; viz: sweet, bittersweet, sharp and bittersharp. Few sharp varieties are now grown specifically because they can be substituted by culinary fruit that is widely available and exhibits a similar flavour balance, and it is the bittersweet group that provides much of the characteristic flavour of English ciders. However it is rare for cider to be made from a single cultivar apple, because the required balance of sugar, acid and tannin is easier to achieve by blending. Other considerations, such as the need for cross-pollination and a spread of harvesting period, also favour the use of a mixed orchard, although cider blending is mainly carried out after the individual types have been fermented.

In the orchard

Many cider orchards are owned or contracted by the large cider companies and this allows the latter to exert control of on-farm activities, although no formal Code of Practice has been drawn up. In general the trees receive a minimum of pesticide sprays once they have been established since the appearance of the fruit is not a controlling factor in their market value, but they do require careful pruning each year. Each season, a small amount of fruit is tested for the presence of pesticide residues and, to date, no instances have been recorded of levels above the maximum residue limit (Loraine Bodington, H P Bulmer Ltd; personal communication).

The raw material that will feed both animals and man
(courtesy: HP Bulmers)

In a traditional orchard of standard trees, the grass sward beneath the trees may be grazed by cattle or sheep in the early part of the season. Although minimum withdrawal periods are observed, wherein animals are excluded from the orchard prior to harvest, this traditional practice would seem to be inconsistent with the objective of avoiding any risk of fruit contamination. This is particularly so when grazing is combined with the traditional harvesting method of shaking the fruit from the tree, and picking them up from the ground by hand. However, the great majority of commercial orchards being planted today consist of bush trees and the grass in these orchards is not grazed. Modern practice typically involves the suppression of grass growth by chemicals and its mowing by machine, but apple harvesting machines still shake the fruit from the trees and fallen fruit are blown or brushed into alleyways for collection by another machine.

In contrast to culinary apples that are harvested early, and ripened off the tree in line with market demands, cider apples are largely ripened on the tree in order to ensure that the fruit is fully ripe before cidermaking. However, this leaves the crop vulnerable to both storm and bird damage (Hulme & Rhodes, 1971). Commonly, the various cider apple varieties are harvested in sequence over the season, but a short storage period may follow in which the starch content of the apple is allowed further time for its conversion into sugar. Storage needs to be closely monitored because the apples must be processed before they begin to deteriorate – soft apples do not press easily and the quality of the juice declines. The harvesting procedure inevitably results in damage to the fruit, and this would appear to preclude a prolonged storage period.

The importance of the natural conversion of starch to sugar can be appreciated when comparison is drawn with the brewing of beer – cider makers have no equivalent to the malting industry that produces enzymes to accelerate the saccharification process.

Cider apples typically contain some 7-9 per cent sugar when ripe – equivalent to 45-60 per cent sugar on a dry matter basis. This range puts the potentially fermentable carbohydrate in cider apples at a lower level than would be found in most malts – information presented in the Brewing and Malting chapter shows a range of sugar plus starch values in various malts of 54.5-72.5 per cent of the dry matter. The pressing of the juice from the apples also leads to a less efficient removal of the sugars than the brewer achieves with a hot water extraction procedure. Thus the sugar content of the expelled apple residue (known as the pomace) constitutes some 10-14 per cent of the dry matter, whereas the average starch plus sugar content of spent grains is only 7-8 per cent. Although such figures confirm that the cider maker derives less fermentable carbohydrate than the brewer from each tonne of raw material, the advantage swings the other way when the co-products are compared - the less sugar that is removed from the apples the more that remains in the pressed pulp. However, the sugar content of fresh pomace is not sustained from factory to feeding trough, because fermentation of this constituent occurs readily in the co-product, and a high sugar level may become replaced by a high alcohol content (Alibes *et al.*, 1984).

Some orchards are equipped with apple washing equipment so that the fruit can be cleaned prior to transport, and this relieves pressure on the washing facilities at the factory. Many factories reserve the right to reject loads that contain an undue proportion of orchard debris, and the trend towards cleaning prior to dispatch seems likely to continue. The supply of fruit to the cider factory is carefully scheduled and hauliers provide a regular transport service during the season. However, there is no formal requirement that vehicles comply with a Code of Practice that would impose cleaning routines and exclude the carriage of non-compatible loads. Purchasing contracts typically state, somewhat briefly, that an open container used for fruit delivery must be in a clean condition and free from contamination. Some fruit continues to be delivered in bags, and here the requirement may seem unduly lax compared with the carriage of other food materials... "bags which have previously been used for artificial manures and/or other objectionable materials must be thoroughly cleaned before use". It would seem to be a small but valuable step in the direction of food safety if the industry were to adopt, for its apple transport, the minimum standards of the BFBi Code (BFBi, 2000) that are applied to the carriage of its moist co-products back to farm. This Code, which includes specific reference to apple pomace, defines cleaning (and special cleaning) procedures and it proscribes the carriage of a number of materials by vehicles that will subsequently be used to carry feed materials.

At the factory

At the factory, the apples are tipped into what are known as canals, though they more closely resemble silos with steeply sloping concrete walls. In the base, a water-filled channel carries the apples into the inspection area. Some orchard debris is removed as the apples are conveyed into the canals, but it is primarily in the flumes that the apples are washed and separated from grass, twigs, leaves and stones, and the circulating water is continuously strained. The silos and flumes themselves are subjected to a washing routine involving high-pressure water jets and, once the apples have entered the factory, the selected fruit will be washed again by a series of spray washers.

British cider makers use more than 100,000 tonnes of apples each year

(courtesy: HP Bulmers)

The washed fruit is then inspected and any damaged or diseased specimens are rejected and consigned to landfill. Selected fruit is passed to a rotary mill or hammermill – depending on the type of press that will be used - where sharp knives cut the apples into small pieces, some 2-8 mm across, in preparation for the pressing that will liberate the juice. Apple presses come in a variety of forms, but the larger factories now use either a continuous belt press that carries the milled apple between rollers, or a hydraulically controlled piston press that squeezes the fruit in a series of tubes and expels the juice through open mesh nylon sleeves. The choice of press has implications for the quality of both the cider and the pomace - belt presses have a higher throughput, but their yield of juice is typically lower than that of a hydraulic press. The belt press also produces a juice with a higher content of fruit solids, and this combination of a smaller amount of juice with a higher dry matter content, implies that the converse will apply to the pomace - namely a larger volume of material with a lower dry matter content.

Interestingly, the milled apple has traditionally been referred to by the cider industry as mash or pulp, while the juice-expelled co-product was known as **apple pomace.** This terminology contrasts with that employed in the citrus juice industry, where it is the pressed fruit that is known as the pulp. The term apple pomace itself is tautologous, since "pomace" is defined as the solid residue of apple juicing (Oxford, 1995). Although traditional English terminology is increasingly being brought under the influence of the suppliers of cider-making equipment in continental Europe, the term pomace appears to be well established and widely understood. Regardless of the finer points of etymology, apple pomace is a well-defined material that can be marketed directly as animal feed, or further processed where facilities permit, to recover the significant pectin fraction.

Before the separation of the pressed solids into the co-product stream, further juice is obtained by the addition of a small volume of water. Typically, sufficient water is added to increase the volume of pressed pulp by about 10 per cent, and the mix is then allowed to stand for about an hour prior to a second pressing operation. Pectinase and/or cellulase enzymes may be added to help degrade the cell walls and this has been shown to facilitate juice removal (R. Peers, Bulmers; personal communication). This technique of double pressing has been employed for centuries and, traditionally, the juice from the second pressing was kept separate and used for the production of a weak cider that would be handed out to workers in part-payment for their services. It was usually referred to as "haymaking cider". The pomace that becomes available as animal feed in Britain has mainly been double pressed, and this may represent a source of variation between British and other apple pomaces, since the solids removed in the second pressing operation may be expected to deplete the pomace of digestible carbohydrate.

Fermentation

When pressed, fresh juice contains a large number of different species of yeast and bacteria and, if left at ambient temperature above 10°C, fermentation will start naturally and lead to the development of a range of distinctive flavours and aromas (Blair, 2001). Traditional cider makers rely completely on this natural process and no external source of yeast is added. The major UK cider makers aim for a more controlled fermentation; they pasteurise the juice and introduce specific strains of Saccharomyces. Typically, they will use some of the yeast strains that are employed in the fermentation of white wines, and a mixed inoculum of S. uvarum and bayanus is quite common. They may be supplied in dried form and simply hydrated in warm water before being pitched into the juice, although some cider factories prefer to propagate their own strain of yeast (Carr, 1970).

Prior to fermentation, the juice may be concentrated and stored at low temperature for subsequent use. But, if storage is not required, the juice will be transferred immediately to a fermentation vessel. Stainless steel, glass or food-grade plastics have largely replaced the traditional oak fermenters, because these materials are easier to clean, and Good Manufacturing Practice requires both vessel and pipework to be sterilised on a regular basis. As well as the yeast, other nutrients need to be added to support yeast growth – apple juice is notably deficient in crude protein and requires supplementation. Additional amounts of the B vitamins, thiamine, pantothenic acid, pyridoxine and biotin may also be required, and a commercial supplement that also includes a range of trace elements is sometimes used to ensure good yeast nutrition. The enhancement of these micro-nutrient levels is particularly important if pure sugars are used as adjuncts to the apple juice - glucose, fructose, and sucrose are all used on occasion. A source of insoluble solids – typically 0.5 per cent bentonite – may also be added as an inert base on which the yeast cells might grow and liberate the alcohol and carbon dioxide that are the end-points of their activity (Carr, 1970).

The pH of the fermenting liquor is adjusted to a level of 3.8 or below, usually by the addition of lactic or possibly malic or citric acid, and sulphur dioxide is used to control the microflora. Enzymes may be used to reduce the propensity of pectin and starch residues to develop a haze or to cause filtration problems.

A rapid and complete fermentation is the objective of British factory cidermaking, and this is encouraged by maintaining the fermenter at a temperature of 15-20°C for approximately two weeks, during which time the alcohol content increases to 10-12 per cent. Although this fermentation is much faster than would be typical of the cool, slow process used in France, it is much more prolonged than the two to three days for which beer is brewed. This, together with the high alcohol concentration of the final product, explains why despite a five-fold increase in biomass the yeast is unsuitable for recovery and re-use within the factory. Two days after the fermentation is complete the yeast is no longer viable and it tends to flocculate and settle in the bottom of the fermenter. Known as cider lees, this co-product is drawn off and, after filtration to recover the cider, it is made available as an animal feed. The rest of the cider will also be filtered and any suspended solids will be added to the lees.

The lees obtained from this process are distinctly different from the liquid produced under the same name in a distillery. Whereas the distillers' lees contains no suspended solids and a very small content of soluble material, cider lees contains all the non-fermented suspended solids, and the valuable yeast fraction, from the fermentation vat. The volume of cider lees produced by each factory may not justify an investment in further processing and, although the material is filtered it is not pressed, and consequently, the dry matter content may be relatively low. This will impact on the cost of transporting the co-product to the point where it will be used as an animal feed, but the nutrient content of this yeast-enhanced material could make the effort worthwhile. Factual information about cider lees appears to be lacking, and thus its potential cannot be assessed.

Further processing of the pomace

Reference has already been made to the extraction of pectin from apple pomace, and to the co-production of pectin-extracted fruit. At the present time, pectin extraction is carried out in only a very small number of factories around the world, and the pectin-extracted co-product is often derived from a mix of both apples and citrus fruit – largely limes and lemons.

Apple pomace has also been used – at least experimentally - as a substrate for the production of industrial alcohol, making use of the co-product's residual sugar content (Joshi and Sandhu, 1996). The alcohol-extracted solids were then considered to be potentially suitable as animal feed. However, the removal of the prime source of digestible energy from apple pomace seems likely to have a significant negative effect on nutritive value of the residual material.

Co-products of apple processing

Apple pomace

Apple pomace is the solid co-product that remains after apples have been pressed to release the juice in cider and apple juice manufacture. It comprises the peel, flesh, core and seeds of the fruit and is a material that is rich in both soluble and insoluble fibre, but one that is low in protein and minerals. The fibre contains a relatively high lignin content – Givens and Barber (1987) found 15-18 per cent (permanganate) lignin as part of the 42-56 per cent NDF fraction in the dry matter of five samples of apple pomace. This degree of lignification has negative implications for the digestibility of the feed, and Givens and Barber found digestibility values for the energy and the organic matter of apple pomace to average only 52 and 58 per cent respectively. However, there was considerable variation between the samples in their study, and the organic matter digestibility ranged from 49 to 68 per cent. Substantially higher organic matter digestibility values of 74-80 per cent were measured by Alibes *et al.* (1984) in three trials with ensiled apple pomace in Spain. This material had a lower fibre content (NDF 42 per cent) than those tested by Givens and Barber, and it was derived from fresh pomace, which had an even lower fibre level (NDF 34 per cent). This appears to indicate that the Spanish investigation involved a different type of apple pomace – possibly derived from culinary rather than cider apples.

The crude protein content of apples is very low (~4 per cent of the dry matter, MAFF, 1990) and the protein level in the pomace is also quite modest (4.4-4.9 per cent, Alibes *et al.*,; 3.0-6.7 per cent, Boucqué and Fiems 1988; 6.2-7.9 per cent, Givens and Barber). The nutritional value of this protein for ruminant animals is also very limited, and restricted protein bio-availability may be associated with the high tannin content (Mueller-Harvey, 1999), for which cider apples are positively selected. Givens and Barber (1987) found protein digestibility values ranging from –24 to +29 per cent, while Gasa *et al.* (1992) found that rumen ammonia levels were insufficient to achieve optimum rumen degradation when ensiled apple pomace was fed as the sole diet. It seems probable that apple pomace will make very little contribution towards meeting the protein needs of farm livestock.

The co-product does have some potential as an energy feed, though the high lignin content indicates that it is likely to be more suitable as a feed for ruminant rather than non-ruminant animals. NFE levels in apple pomace are quite high; 61.8-70.6 in Givens and Barber's study, and an estimated 75 per cent for the fresh pomace used by Alibes *et al.* However, this fraction was not fully defined in either study.

Apple pomace is a palatable feed
(courtesy: HP Bulmers)

Apart from insoluble fibre components, the largest constituent seems likely to be pectin at approximately 15 per cent of the dry matter, followed by sugars, which may vary between 10 and 14 per cent. Fresh apple pomace may also contain a proportion of starch (2-8 per cent), and a significant content of both alcohol and organic acids – the latter giving the fresh co-product a pH value of 3.5. All of these fractions may be expected to be highly digestible – in contrast to the poor digestibility of crude fibre, which Givens and Barber found to average only 46 per cent, and of crude protein, which these authors found to be typically negative. Such wide disparity between the digestibility of different fractions highlights the importance of determining the chemical composition of material from a particular source, since any difference in composition is likely to have a significant effect on feeding value.

Givens and Barber determined in vivo ME values with sheep and found values ranging from 7.5 to 10.2 MJ per kg DM. These results indicate that the average material has an energy value similar to that of grass hay, but the wide variation appears to confirm the need for samples to be individually assessed. However, the GE values in Givens and Barber's study were determined by bomb calorimetry on the dried pomace. Although the loss of a volatile organic acid content of perhaps two per cent on oven-drying would enhance the GE of the remaining feed marginally, the effect of the loss of alcohols would be more substantial. If the fresh pomace had contained 5.6 per cent alcohol, as measured by Alibes *et al.* (1984) in their material, its loss on oven-drying would reduce the GE of the remaining feed by approximately 0.7 MJ/kg dry matter. It may also be notable that the Givens and Barber evaluation included a prediction of methane losses by the Blaxter and Clapperton (1965) equation. For other fibrous feeds this equation has been shown to overestimate the methane loss - e.g. brewers' grains (Rowett Research Institute 1984) – and if a similar overestimate has been applied to apple pomace, the consequence would be a reduction in the determined ME value by some 0.8 MJ/kg dry matter. Adjustment of the ME values determined by Givens and Barber to take account of both these factors could increase the average ME value to 10.2, and the range to 9.0-11.7 MJ/kg DM.

In most practical situations, apple pomace will be stored before feeding and this may be associated with substantial changes in feeding value. Alibes *et al.* (1984) noted that ensilage brought about a significant release of an effluent rich in digestible nutrients, and the production of apple pomace silage that contained a higher concentration of both protein and fibre. The organic acid content of the silage also increased substantially, including increases in lactic, acetic, butyric and isovaleric acids. But the most dramatic effect, of a fermentation that involved yeasts as well as bacteria, was a large increase in the alcohol content – the total alcohol level increased from 5.6 to 18.9 per cent of the dry matter. Such a high alcohol level in the silage would appear to imply an unusually high concentration of fermentable sugars in the pomace, which would represent inefficient processing by the apple presses. N.B. The alcohol level would have been even higher if the effluent had been retained, since the alcohol content of the effluent was richer than that of the pomace silage. If this finding were typical of pomace ensilage, there may be implications for animal health associated with the prolonged use of this alcoholic feed. Alibes and his colleagues merely reported that, in their trials, the animals on the pomace diets required "a considerably longer resting period".

Apple pomace has supplied up to 50 per cent of the dietary dry matter for pregnant sheep without adverse effects on intake and performance (Rumsey and Lindahl, 1982). A higher inclusion rate has been used successfully, for both sheep and beef cattle, when adequate protein supplementation was given (Alibes *et al.*, 1979, Thonney *et al.*, 1986). High levels of apple pomace have been associated with a significant reduction in rumen pH (Manterola *et al.*, 1999), and the need to balance this effect should be considered when formulating the diet. A number of studies have shown poorer performance when the co-product was fed in conjunction with non-protein nitrogen sources (Fontenot *et al*,. 1977, Oltjen *et al.*, 1977). Apple pomace has also been used as a supplement to pasture for grazing dairy cows, when it was associated with an increase in milk output (Edwards and Parker, 1995). In this New Zealand study, it was noted that the pasture substitution effect of the pomace was small, and thus the supplement resulted in a significant boost to feed intake.

3 Bread and Baking

Bread is believed to be the oldest manufactured food, having been produced in unleavened form in the Middle East some 10,000 years ago. At that time, a mixed variety of grain was employed for the purpose and it was crushed by hand using a pestle and mortar. The skills of bread making were developed by the Ancient Egyptians, who were claimed to have used tougher wheat varieties – in terms of the elasticity of the dough - and harnessed the abilities of wild yeasts to lighten the finished product. The Egyptians also invented the closed oven in which the bread was baked and the product assumed great significance in their culture, sometimes being used in place of money. During this period iron ploughshares were introduced, oxen were yoked to pull them and, with these significant advances, bread became adopted as a staple food throughout Europe (FOB).

It was the Romans who introduced consumer choice; wholemeal bread may have been suitable for the masses, but the wealthy classes expressed a preference for the more refined and expensive white bread - made from flour from which the bran had been removed. That preference continues to be expressed by many of today's consumers, though the distinction in price appears to have been reversed. It was only in the 1700's that wheat began to overtake barley as the principal grain for bread making, but by 1826 the military authorities were recommending the wholemeal type as being healthier – though their evidence for this claim has not been documented.

History has not been kind to the miller and the baker, who appear to have had an enduring ability to inspire distrust. In 1202 a law was introduced to regulate the price of bread but, by 1266, it was found necessary to create the Assize of Bread which linked the price to the weight of the loaf. Anyone breaking this law could be pilloried and banned from baking bread for the rest of his life. However, in Chaucer's 'Miller's Tale', written in the 14th century, the miller continued to be identified as "greedy" and the suspicion with which he was widely regarded was noted. Again in the 1700's still further laws were enacted, aimed at controlling the amount of bread supplied for a given price but, a hundred years later, a wide section of the populace was starving and the price of bread was out-of-reach of the common man. By that time the standard weight had been abolished, and bakers were required to prove the weight of their loaves at the point of purchase. But if that solved the quantitative dispute, it was not long before the bakers were back in the dock with qualitative complaints – less reputable members had found that bread could be whitened by adulteration with lime, alum and chalk.

That chequered career of the millers and bakers may partly explain the close interest that Government bodies have traditionally shown in bread manufacture, but their continuing interest is more likely to be due to the key position that bread has established in the national diet. Recent data on food eaten in the home show that bread provides about 14 per cent of both energy and protein consumption (Flour Advisory Bureau), and nutritional guidelines for health education include a recommendation that bread consumption should be increased (NACNE, 1983; COMA, 1991). Great emphasis is now placed on the value of dietary fibre, and this appears to confirm the wisdom that lay behind the military authorities' preference for wholemeal bread during the nineteenth century. At 8.5 per cent dietary fibre (on a fresh weight basis), the wholemeal product contains more than three times the fibre level found in white bread. Since 1941, other nutrients have been incorporated into the flour by Government edict, in order to enhance its value as a supplier of vital micro-nutrients. Chalk is no longer considered to be an adulterant but a calcium supplement, and non-wholemeal flours are also fortified with iron, thiamin and niacin.

Food safety issues

The use of bread as a vehicle for improving the human diet may be taken to imply a degree of confidence that any food safety issue has been satisfactorily decided. However, in the 11th millennium of bread production, millers are aiming to provide even greater assurance to their customers and consumers; "not only is their product safe, it has been produced from wheat that has been grown and stored in accordance with the best farming practices". In partnership with the National Farmers Union, and others in the cereal sector, the millers' representative body, nabim, has developed the Assured Combinable Crops Scheme (ACCS). This scheme sets out clear requirements with respect to the practices employed in the production and storage of grain on farm, and a separate scheme, known as TASCC, extends the same degree of care into off-farm grain storage and delivery. Together these schemes produce a complete chain of assurance from the field to the flour miller's reception point. From September 2000 these schemes became a commercial reality for many UK millers, who undertook to source grain only from assured suppliers or from any others who could show independent verification of equivalent standards. The flour millers firmly believe that membership of schemes such as ACCS will ensure better standards of hygiene, better understanding of food safety requirements, and a willingness on the part of wheat producers to demonstrate that the high standards of the scheme have been put into practice. N.B. The millers' participation in such schemes guarantees the quality of the flour that is used in the production of other bakery items, as well as bread.

Wheat is obtained only from assured farms
(courtesy: Roquette UK)

ACCS and its Scottish equivalent, SQC, require their members to be aware of and to comply with a whole raft of guidance on the safe use of pesticides, fertilizers, grain store management, salmonella control and the further requirements of the Food Safety Act 1990. Safety issues are linked with environmental protection and ACCS members must also comply with the requirements of MAFF Codes for the protection of soil, air and water. Comprehensive records are needed to provide the evidence that will be checked by a team of verifiers, who will seek to confirm that only approved chemicals were used by certificated staff in the appropriate amounts and in suitable weather conditions. Assurance must also be provided about the grain itself; that labelled crops were put into clean stores, protected from insects and rodents, and the moisture content and temperature monitored regularly using calibrated equipment. As a final measure to ensure the traceability of grain supplies, growers will have retained samples of the wheat that was moved off-farm, accompanied by a grain passport, in clean vehicles that were operated according to the UKASTA Code of Practice for Road Haulage.

Millers have long held a positive view with respect to pesticides - that such chemicals were used by wheat growers in ways that conferred a benefit on the product (nabim 1998). By improving the plant's defence against insect and fungal attack – including the toxin-producing ergot - appropriate pesticides reduce crop damage and diminish the consequent effects on both the yield and quality of the grain. Pesticide use after harvest is also recognised as a storage aid that offers protection for consumers from the contamination associated with insect infestation. But the industry is aware of consumer concerns about pesticides in food, and members have adopted a responsible approach to ensure that any pesticide residues are kept well short of the level at which they may constitute a problem. Thus the flour milling industry has been monitoring pesticide residues for more than 20 years, by means of the independent testing of both imported and home-grown wheat grain that arrives at the mills. The results have been consistently reassuring; the great majority of samples examined contained no detectable residues, and only a tiny proportion showed residues above the MRL's that reflect the limits of good agricultural practice. Table 1 shows the nabim pesticide residue data for the last seven years – earlier results appear to be unavailable.

Table 1: Summary of results of nabim pesticide monitoring 1994-2000

Year	*1994*	*1995*	*1996*	*1997*	*1998*	*1999*	*2000*
Samples analysed	687	422	331	328	178	172	190
Samples negative* %	76.1	77.7	75.2	72.9	83.1	90.7	82.0
Samples above MRL %	0	0	1.2	0	0	0	0

Source: nabim, personal communication
*Reporting limit 0.05 mg/kg, equivalent to 1% of the MRL or lower.

N.B. Although the number of samples has been reduced in recent years, the survey covers all nabim members and individual samples are taken in proportion to wheat usage. The picture is thus considered to be representative of wheat usage from all sources.

The current industry survey covers only organophosphorus chemicals used in grain stores. Earlier surveys included organochlorine compounds, synthetic pyrethroids, methyl bromide and other compounds used in field or storage applications. These chemicals were rarely if ever found and have thus been dropped from the routine tests. This may be taken to imply that the achievement of negative finding in more than 80% of the current tests represents an improvement on a similar finding in earlier years. Routine tests are still supplemented by a number of spot checks.

Figure 1 contrasts the pesticide residue levels of wheat obtained from UK and imported sources over the last seven years. Home-grown wheat samples have shown a consistently lower proportion of positive results, and this may reflect greater attention by UK farmers to careful and well-timed applications.

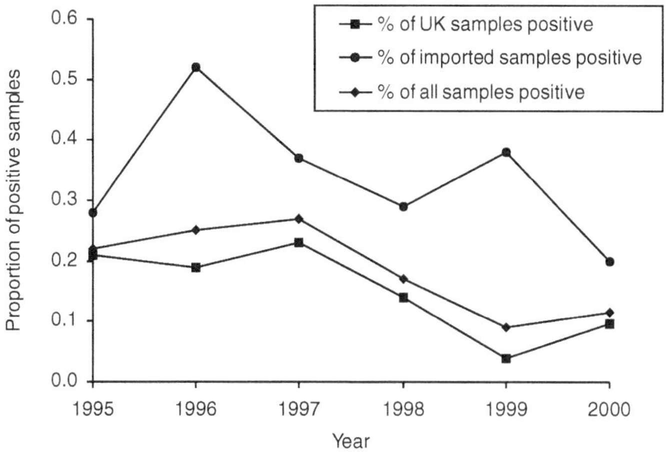

Figure 1: *Comparison of pesticide residue findings in UK and imported wheat 1995-2000 (Source: nabim)*

Individual millers also carry out their own pesticide checks, and cross-check the results with the declarations made on grain passports - any discrepancies being investigated with the grain supplier. Restricting their wheat supply to ACCS members only, will enable millers to provide well-documented assurance to the bakers and strengthen the guarantee that has previously been provided in the grain passport. At the time of its introduction, the grain passport was itself believed to be unique – providing un-matched protection against unnecessary and multiple treatment of grain. Taken together, it may be expected that the additional measures will lead to a reduction in the pesticide residues occurring in flour and build increasing consumer confidence in bakery products.

Another aspect of quality that has been of particular concern to consumers in recent times is the genetic modification of certain crops. This concern has had a major impact on maize processing, but the milling and baking industries have been able to avoid the controversy because no genetically modified wheats have been commercially available in the UK. Research developments indicate that the first GM varieties may be available for commercial production in Canada by 2002, but that is not to say that they will be grown or imported. The UK flour milling industry is sensitive to consumer concerns on this subject, and its policy on biotechnology was drawn up as far back as 1994. "Flour millers would need to be convinced of the safety and public acceptability of GM technology, and the practical potential of GM wheat varieties, before they would be brought into use" (nabim, personal communication; February 2001).

Bread

The basic ingredients of bread are flour, yeast, salt and water and these can give rise to a wide range of different types of bread. The addition of other minor ingredients such as sugar, fat, milk, malt and eggs can extend the range still further. Quantitatively, flour is the principal constituent and its composition varies according to the extraction rate of the wheat. Wholemeal flour, as the name implies, represents a 100 per cent extraction of the original grain, whereas brown and wheatmeal flours contain 85-90 per cent and white flour only 72-74 per cent. The excluded fractions originally made up the outer layers of the grain, and compared to the extracted flour they contain greater concentrations of protein, fibre and mineral elements. Much of this material, comprising a spectrum of milling co-products from bran to wheat middlings, is directed to the animal feed market where it is valued both for its nutrient content and for its water-binding and generally stabilising effect on the gut.

Bread is produced by baking a dough made from a mixture of wheat flour and water (approximately 1 part flour to 0.6 parts water), that has been aerated with carbon dioxide produced by yeast fermentation. A small amount of alcohol is also produced by fermentation but this is subsequently driven off during the baking process. Up to 2 per cent salt is added to give flavour and, because this may depress yeast activity, the addition of salt is sometimes delayed to a later stage. Because of the key role bread plays in the human diet, the baking industry has been obliged to take account of an additional factor. A high salt intake has been identified as a primary cause of hypertension in human populations (WHO, 1982), with a consequently increased risk of ischaemic heart disease. This potential risk has led to pressure from human nutritionists for bakers to reduce the salt content of bread as a positive benefit to the whole diet. The industry has responded to such urgings and, since the late 1980's, the salt content of bread has been reduced in two steps by more than 20 per cent (J White, Federation of Bakers; personal communication).

The yeast inclusion rate is also varied, between 1 and 2 per cent, in line with variations in the activity of the chosen yeast strain and the amount of time that will be allowed for fermentation. Much of the bread made in both large and small-scale bakeries in the UK is produced by the Chorleywood Bread Process

(Pickles, 1968), in which fermentation is supplemented and partially replaced by a large amount of mechanical processing of the dough. More active yeast strains have been developed for this process and additional amounts are used, together with a small percentage of high-melting point fat (about 0.1%), which improves the gas retention properties of the dough. Ascorbic acid (vitamin C) is also added to the dough and this, after conversion to the dehydro form, helps to form disulphide linkages which give improved structure to the bread. A small amount of mould inhibitor – usually acetic acid and sometimes calcium propionate – may be added to the dough. Though this may help to safeguard the dough from microbial attack, the benefit of any volatile preservative seems likely to be diminished during the baking process – leaving the bread a potentially more vulnerable material.

There follows a series of steps in which the dough is divided into pieces of equal volume, rested and then moulded to meet the specification before being finally proved – another holding period in which the pieces of dough are subjected to a warm moist atmosphere for 45-50 minutes. Limited amounts of dough become available for animal feed – usually when the production line is being changed from white to brown or to wholemeal dough, and there is an inevitable mixing of the different materials. Individually, each one represents a satisfactory, food quality material, but in combined form they are unsuitable for defined food markets. For the animal feed market, such minor variations are of little significance. Dough may be expected to have a very similar composition to bread; particularly since much of it is cooked before it is consigned to the feed market.

After proving, the dough is typically baked at an oven temperature of 230°C for 23-30 minutes. This results in a core temperature of some 96-97°C at the point when the loaf leaves the oven, although baking may continue for some minutes as some of the surface heat is transferred to the core. At this point, the crust is virtually completely dry and it acts as an effective seal for the remainder of the loaf. But this situation does not last; the loaves are immediately dislodged from their tins and left to cool in a humidity-controlled atmosphere for 2-3 hours and, during this time, some of the inner moisture transfers to the crust. Vacuum cooling is employed in some bakeries and this can provide a useful shortening of the time during which the soft, warm product is vulnerable to both physical damage and microbiological contamination. Much of the cooled bread is then sliced, and this destroys the protection that may have been afforded by the crust. However, sliced loaves are typically packed in waxed paper, heat-sealable film or plastic bags and this offers an alternative form of protection.

A proportion of bread is rejected at the bakery. It may fail to satisfy a range of quality assurance checks in relation to the specification of the product, but often it is rejected because it is under or over the prescribed pack weight although otherwise perfectly satisfactory. Such fresh, out-of-spec food-quality product is immediately diverted to the animal feed outlet. Feed bread is also collected from supermarkets on a daily basis. Although the shelf-life of different breads may vary from hours to a matter of days, all of them are typically regarded as unsaleable after 24 hours on the supermarket shelf. Beyond that deadline, the bread is removed from the food supply and redirected to the feed chain.

Within two days, some of this bread will be on-farm
(courtesy: Federation of Bakers)

Although identical in composition to food quality material, this feed bread cannot be used without further processing to remove its wrappings. Typically, it will be collected in sealed, compactor vehicles and taken to a shredding plant, where any metal ties will be removed and the light-weight shredded wrappers will be stripped from the bread and aspirated away. The bread is then pushed through a fine sieve, which excludes further pieces of wrapping and, beyond this point, it is claimed that "virtually all" the wrapping has been removed by the processing – though close examination may reveal a small proportion of finely-shredded pieces

(P Latham, SugaRich; personal communication). The waste material is consigned to land-fill while the **processed bread** may be dried or marketed in its fresh form. Dried bread is mainly used as an ingredient for compound feeds, although a proportion is marketed directly to livestock farms. The undried bread is supplied directly to farm for use as a feed for both pigs and cattle.

Flour confectionery

Flour confectionery is a term used to distinguish biscuits, cakes and pastries from sugar confections such as sweets and chocolates. Since the whole group of flour-based products is essentially composed of a small range of highly digestible ingredients – flour, fat and sugar – it is possible to assess the nutritional value of any feed material of this type by reference to the proportions in which these basic elements are used. Table 2 provides a general guide, though further processing – such as the addition of chocolate coatings or the insertion of cream – can make a significant difference to the energy value of the final product. It should be noted that the ingredient proportions in Table 2 are given on a fresh weight basis, and recognised that the unbaked mixes of different types of flour confectionery will include different amounts of liquid. While such differences expand the range of energy values in the unbaked mixes, it may be expected that much of the variation in moisture content will be lost in the oven. For this reason, the energy value of each type of baked product has been calculated to a standard 90 per cent dry matter basis. The energy values are indicative of the range that may be available, though some variation in moisture content may occur and the actual dry matter content of any supply would need to be checked.

Table 2: Proportions of major ingredients in the principal types of flour confectionery

	Puff pastry	*Short pastry*	*Cake incl. fat*	*Scones*	*Fermented e.g. bun loaves*	*Sponge no fat*
Flour	42	52	30	49	51	31
Fat	32	26	18	9	8	
Sugar			22	9	8	31
Eggs			20		3	38
GE value*	24.4	22.4	21.4	18.1	17.7	17.0

Source: Bennion and Bamford (1983)

The GE values have been calculated from "calorific values" for the ingredients given in the Manual of Nutrition (MAFF, 1961), and expressed in MJ/kg on a 90 per cent dry matter basis.

Biscuits

Biscuits are a traditional type of flour confectionery and are baked in traditional manner from a dough that is principally made from flour, sugar and fat. A huge range of biscuit types is manufactured but they are basically made from only two types of dough, hard and soft. Hard dough has a high water content and relatively little fat and sugar; it is used in the manufacture of crackers and semi-sweet biscuits. Soft dough has much higher proportions of fat and sugar, but contains much less water. Soft dough is widely used in the manufacture of a large range of biscuit types and a variety of other ingredients may be included. After baking, secondary processing may add chocolate coatings and additional flavourings, or sandwich the biscuits with fat-based "cream" fillings. By this stage, any difference in the moisture content of the dough will have been lost, since all biscuits have a typical dry matter content in excess of 96 per cent. When packed in moistureproof wrappers, biscuits have a long shelf-life of the order of six months or more in the human food market (Manley, 1998).

Modern biscuit plants tend to operate largely continuous operations and to produce a relatively narrow range of biscuit types in order to minimise the changeovers that would inevitably reduce the daily output. Product changes also result in a higher proportion of biscuits that fail to match the required specification and they have to be redirected to the feed supply. Out-of-spec biscuits are immediately conveyed to the feed bins and there they are added to others that have been broken, mis-shapen, wrongly-labelled or may be the perfect product of new lines not yet released onto the human food market. Such off-the-production-line biscuits are considered to constitute 90 per cent of the **biscuit meal** that becomes available as animal feed (P Latham, SugaRich; personal communication), and they are largely unwrapped. The remaining 10 per cent is made up of wrapped biscuits that have exceeded their sell-by date – though the wrappers will be stripped from the biscuits prior to their placement on the animal feed market.

The term biscuit meal is sometimes used to cover a broader range of bakery blends, and this wider definition may have potentially serious implications if such blends were to contain meat products or even to originate from premises where meat-containing products were produced or prepared. Following the outbreak of foot and mouth disease in Britain during 2001 – a problem that still continues as this book goes to press - and the linking of the initial problem to inadequately-processed food wastes on a swill-feeding unit, the use of catering wastes as feed has been banned. The Animal By-Products (Amendment) (England) Order 2001 came into force on 24 May. Thus, in order to avoid the classification of wholesome and acceptable bakery co-products as wastes that require licensed disposal, it may be advisable for producers of bread, biscuits and pastries to exclude any meat operations from the site. For biscuit manufacturers, it would also seem prudent to reclaim ownership of the biscuit meal term.

For the animal feed market, the major focus of interest is the fat content, since this largely determines the energy value of biscuits, which are essentially an energy-feed. The fat content – always of vegetable origin - varies from 0.2 per cent in crisp-breads to perhaps 25 per cent in "cream" and chocolate-coated sweet biscuits, though most biscuit meals would represent a mixed supply.

Bakery co-products

Author's note: Although the materials described in this chapter are all unquestionably feeds from the food industry, there is a view that they are not co-products. The principal difference that distinguishes these feeds from those described elsewhere in this book is that they are not simply food fractions, such as potato peel or wheat bran, but they consist of a blend of raw materials. In itself, this feature does not set them completely apart, there are combination feeds in other sections such as molassed sugar beet pulp and maize gluten feed. But unlike bakery items, these feeds represent recombinations of fractions that were originally present in the same raw material. However, there are other examples, such as potato crisps and chips, where material of potato origin is brought together with oil from a different source, and the two elements are combined during cooking. From the animal feed viewpoint, bread, cakes and pastries represent a similar meld of ingredients - processing in the bakery has resulted in irrevocable changes such that the components are not available separately. Thus, in the author's view, bakery items are not fundamentally different, and the most common examples are included in this book of co-product feeds. N.B. Part of the argument against the classification of bakery items as co-products is the belief that they are in reality the finished "products" of these factories. The implication of that argument is an acceptance that the classification of feeds should be based on the objectives of other industries. This argument is discussed and refuted in the introduction to this book.

Processed bread

Processed bread is largely "yesterday's bread", removed from the food supply to be replaced by a freshly-baked product, but it also includes a proportion of "today's" production that failed to satisfy QA checks at the bakery. Before it reaches the farm it will have been shredded and sieved to remove virtually all of the wrappings and it may have been dried – from its usual 63-65 per cent to around 90 per cent dry matter. Bread is a highly digestible cooked material that is derived largely from wheat flour. Its protein content largely reflects that of the original flour, though its amino acid balance may have been marginally improved by the activity of the bakers' yeast. The raw material comprises a large number of different bread types, produced from a range of flours of varying fibre content. However, the processing of the co-product leads to a blending of the different types and this lessens the variability. Salt is used in the manufacture of most breads, though the amount has been reduced in recent years. The current level of around 0.9 per cent in the dry matter (equivalent to 0.35 per cent sodium) would not be expected to pose any problem for either ruminant animals or pigs.

Dried bread can be stored like any other dry feed in a clean, dry bunker silo, but the undried material may be stored in this way only for short periods. For longer storage, undried bread is preferably stored anaerobically, after compacting and sheeting to avoid aerobic spoilage.

Bread is a palatable, energy-rich feed that can be offered to both pigs and ruminant animals. Bread is essentially an alternative to cereal grain, though its digestibility may be expected to be a little higher than that of the whole, uncooked wheat grain. Bread has a lower fibre content than the wheat that was originally milled for flour, and additional sources of fibre must be provided in the diet. This is particularly important for ruminant animals where rapid digestion of the cooked starch may otherwise lead to problems of acidosis. Fed in large quantities to dairy cows (~8kg per head per day), bread has been associated with a steep reduction in the butterfat content of milk (R Crawshaw, personal experience), but more modest quantities have proved satisfactory, and may be expected to boost both milk yield and milk protein content. Judicious use of bread in a carefully controlled dairy ration may allow some deliberate manipulation of milk composition, where this is desirable. The high starch content makes bread an ideal feed for beef cattle, though it may be prudent to restrict the feeding rate to a maximum of 50 per cent of the dry matter intake.

In the pig, the low fibre content of bread may favour its replacement of wheat in the ration, since this may increase the uptake of both glucose (Jenkins *et al.*, 1978) and amino acids (Just *et al.*, 1983) from the small intestine. However, a minimum fibre supply should be maintained in the pig diet because of its stabilising effect on the gastro-intestinal microflora, and the buffering effect of fibre through cation exchange (McBurney *et al.*, 1983; van Soest *et al.*, 1984). Commercial recommendation (KW 2000) is to limit bread to no more than 30 per cent of the dry matter intake of growing pigs, although Lucas (1966) was content to recommend 20-35 per cent for 45-67 kg animals and 35-50 per cent bread for pigs of 45-90 kg liveweight. Significantly, Lucas's balancer ration was stipulated to contain 20 per cent of relatively fibrous weatings.

Biscuit meal

Biscuit meal is a high dry matter, energy-rich feed that is largely made up of biscuits taken directly from the production line. A small proportion of the supply represents biscuits that have exceeded their sell-by date, but since they have a long shelf-life when stored in dry conditions, such biscuits still retain their freshness and palatability as an animal feed. Some sources of biscuit meal contain a proportion of other bakery co-products. Providing there is no contamination by meat products, such inclusions should be considered on their merits as alternative energy sources.

For livestock farmers, biscuit meal is a rich energy source, with some variation in energy value in line with the variation in fat content. The nature of the fat varies according to the type of biscuits being manufactured, but all will be of vegetable origin. The fat content ranges from 0.2 per cent in crisp-breads to perhaps 25 per cent in "cream" and chocolate-coated sweet biscuits. However, most biscuit meals would represent a mixed supply and 10 samples analysed by ADAS over the period of a year (Ruth Bishop, personal communication) showed a mean oil content of 19.2 within a range of 10.9 to 24.8 per cent. Given such a wide range of fat content, it would seem essential for anyone considering the use of biscuit meal to establish a typical value for the particular source.

Other than the fat, the principal components of biscuit meal are wheat flour and sucrose. This may be taken as an indication that biscuit meal is a feed of low fibre content and limited protein value, but a rapidly available source of sugar and cooked starch. Biscuit meal may be used for both ruminants and non-ruminant animals, but adequate dietary fibre would need to be provided from other feed sources. It would be prudent to limit biscuit meal inclusion in ruminant rations to perhaps 15 per cent, with particular care needed for dairy cows where the fat content of the whole diet would need to be closely controlled. Greater amounts can be used in pig rations; as much as 30 per cent biscuit meal should be possible for sows and 20-30 per cent for fattening pigs, providing the oil content of the diet can be kept within reasonable bounds.

Biscuit meal has a poorer shelf-life than the biscuits manufactured on the same production line that were sealed immediately in airtight wrappers. Biscuit meal's greater surface area means an increased exposure to air and, in time, the relatively high fat content may become rancid. While the co-product may still remain palatable, the development of rancidity would increase the antioxidant requirements of both ruminant and non-ruminant animals, and additional vitamin

E supplementation may be necessary (ARC 1980 and 1981). Biscuit meal is also hygroscopic, which increases the risk that pockets of mould may develop where the previously high dry matter material has picked up moisture from the air. For both these reasons, biscuit meal should be regarded as a feed that needs to be stored well and used quickly, with thorough cleaning between loads.

Breakfast cereals

Breakfast cereals are a dry, high energy, cooked feed that is available as a coarse meal for both the ruminant and non-ruminant markets. This bakery co-product includes a mixture of several types of breakfast cereals that are taken fresh from various parts of the production line. Since the processing of some cereals includes both soaking and moist heat treatment, the co-product material taken from the early stages of the process needs to be re-dried. At the far end of the production line, some of the co-product has been packaged and these wrappers need to be removed. Thus, prior to being made available to the animal feed market, it is common for all of the breakfast cereal to be subjected to both dryng and package stripping processes.

Although there may be minor inclusions of dried fruit, chocolate and nuts, the bulk of the material is made up of cooked cereals and sugar, and this low-fibre feed is highly digestible. Breakfast cereals are also highly palatable and they are used in creep feed formulations. For ruminant animals, the feed can be used as part of a high density ration for high yielding dairy cows and fast-growing beef cattle, or for ewes in late pregnancy. The high content of sugar and cooked starch will be rapidly degraded in the rumen, and breakfast cereals are probably best used as part of a mixed ration, with adequate fibre provided from other sources. For pigs, breakfast cereals represent a high energy feed that should be capable of use at a relatively high level. However, there is limited recorded experience of breakfast cereals in pig diets, and it may therefore be prudent to restrict the inclusion rate to 20 per cent, building gradually to perhaps 30 per cent in a well-balanced ration with adequate dietary fibre.

Cake

The cake that is available as animal feed may have a wide range of composition, in line with that available on the human food market. If it is supplied in fresh form, the dry matter content may vary from 65 to 87 per cent (McCance and

Widdowson, 1991), and the protein content may vary too, though most values are predictably low at 5-7 per cent of the dry matter. All forms of cake represent concentrated sources of energy, and the greatest difference is between those rich in fat and those with a high sugar content. Swiss rolls and fatless, jam-filled sponges may have fat levels as low as 6 per cent, but these types tend to have the highest sugar content, at up to 63 per cent of the dry matter. Cream-filled sponges, and particularly those with butter icing, have a much higher fat content at up to 35 per cent of the dry matter. A commercial supply will inevitably comprise a mixture of different types of cake, but the wide range of sugar and fat values makes it imperative that typical values be established for a particular source.

Cake is a highly palatable feed that is almost completely digestible. It can be fed to both pigs and ruminant animals, providing it is used in a well-balanced ration that compensates for its lack of fibre and pays appropriate attention to the fat and sugar contents. The sugar content of some types of cake is in line with that found in molasses, and it would be prudent to restrict its inclusion in ruminant rations to about 12 per cent. Other cakes have a particularly high oil content and care should be taken to control the fat content of the whole diet to 6-8 per cent. At the higher end of that range, dairy cows may be expected to show a negative effect on the butterfat content of their milk. Higher inclusion rates – up to 20 per cent - could be used for sows, for whom fat is a valuable energy source. Cake could also be fed to fattening pigs, though it would need to be restricted to about 12 per cent in order to moderate the effect of a high fat feed material on the fat content of the pig.

Good, indoor storage facilities are required in order to maintain the palatability and attractive smell of cake but, unlike biscuit meal, cake tends to become hard on keeping. However, the high fat content of both materials is vulnerable to oxidation and may become rancid. Such feeds may still be palatable though, when used, it is advisable to provide increased supplements of vitamin E.

4 Brewing and Malting

The production of fermented drinks is a practice with a very long history. The earliest records point to the Sumerians being adept at the art some five thousand years ago, but there is evidence from Africa, China and both North and South America that brewing was also being practised in those parts of the world during their pre-history (Glover, 1997). In Britain, ale was being brewed when the Romans arrived and still being brewed a millennium later when the Normans invaded. Despite efforts to introduce wine by both of these conquering forces, the northern climate favoured the cereal crop rather than the grape vine, and the British continued their traditional brewing practice. Formal recognition of its place at the core of British life came with the imposition of taxes, first at a local level, and then in 1188 a national tax was levied, ostensibly to pay for the Crusades. Some would complain that the "Saladin Tithe", as it was known at the time, is still being paid more than eight hundred years later! Prior to the fifteenth century, ale was flavoured with herbs but it was prone to spoilage during the summer months. When hop flavouring was introduced in the 1400's, the product's mild antiseptic properties were found to offer improved keeping quality.

Before the 1880's when brewing was established in Britain on a factory scale, much beer was brewed on farms and in monasteries, where a complete cycle of activities was completed. These brewers farmed the land and grew at least some of their barley requirement, they malted the grain and brewed the beer, and then fed the residues to their own livestock. In effect, they operated an integrated system with the full transparency that is demanded by today's animal feed market. In modern Britain, few breweries retain any formal link with either arable or livestock farms, and it is rare for them to own their own maltings – today's malt is mostly produced by specialist maltsters. Each year, more than 1.8 million tonnes of barley are turned into malt and malted barley remains the predominant ingredient of the British brewing process, as can be seen in Table 1.

Table 1: Materials used in Brewing ('000 tonnes)

Malt	683
Other cereals	70
Sugar	114
Hops*	5

*Hops equivalent figure, which includes other processed forms of hops.
Source: BLRA Handbook (2000)

Beer production statistics for the last hundred years (BLRA 2000) show remarkably little change overall, peaking in the late 1970's but falling by some 12 per cent over the last twenty years, to a level that was typical of the period between 1905 and 1920 – though in those early days the figures did include brewing in Southern Ireland. Consequently, the output of brewing co-products has been similarly stable over this long period, and the brewing industry has remained a major producer of moist co-product feeds. Their widespread availability, and popularity with dairy farmers in particular, resulted in substantial numbers of dairy farms coming to rely on these co-products as part of their basal ration. Hay and grains formed the basis of many dairy rations, and the grains were widely considered to add succulence and palatability that helped to compensate for what was often variable quality hay.

Brewers often use a rule of thumb guide to the production of animal feed (spent grains in their terminology), viz: one tonne of malt yields 1.1 tonnes of grains. The application of this relationship to the malt volume given in Table 1 suggests a moist grains output of about 750,000 tonnes per year. A summary of beer production statistics since 1898 is given in Table 2, together with an estimate of the associated output of animal feed. The beer statistics are derived from Customs & Excise records, whereas the feed estimate is based on a broad relationship between beer output and estimated malt input. Although this estimate is a less accurate guide, it suggests that the UK brewing industry has contributed 600-800,000 tonnes of animal feed per year for the last hundred years.

Malting

Careful control of cereal processing from harvest to the final product is recognised as the key to good quality in the brewing industry. A specific example within this broad objective was given by Lloyd (1986), who recommended zero tolerance of mouldy grain; "the removal of mould from contaminated grain is beneficial but insufficient" he pointed out "to ensure that mycotoxins, which have diffused into

Table 2: Beer Production and Estimated Feed Output in the U.K.

Year	*'000 barrels*†	*Moist feed '000 tonnes*‡
1899/00*	37,105	738
1909/10*	34,300	702
1919/20*	35,048	712
1929/30	25,062	586
1939/40	25,367	590
1949/50	26,514	604
1959/60	26,115	599
1969/70	32,940	685
1979/80	41,701	796
1989/90	37,400	741
1999/00	35,843	732

*Early figures include the whole of Ireland
Source: †BLRA 2000
(‡Calculated from estimated malt use and excludes yeast)

the grain, have been completely eliminated". Thus, when barley reaches the maltster the grain is checked for a wide range of potential contaminants, including botanical, entomological, microbiological, chemical and physiological defects as well as discolouration and inorganic components. Good quality grain contains no ergot or evidence of insect infestation, it is not mouldy or mixed with other cereal grains, nor damaged or discoloured and the grain is not germinated or above a specified nitrogen content; it is not contaminated with stones or lead shot. For the maltster, a crucial test is of "germinative capacity" – although some newly-harvested grain may be dormant for a period after harvest, the maltster needs assurance that, when it is used, germination will exceed 95 per cent.

If it passes such scrutiny, the barley is then screened to remove the awns and small grains, and dust particles are aspirated away. There may also be a proportion of chaff if the feed has been inadequately cleaned prior to delivery. The material passing through the screens is marketed to the animal feed market as **barley screenings** – sometimes in conjunction with the **malt powder** that is collected from various points throughout the operation and malt screenings obtained at the end. As a feed, barley screenings are relatively dry but variable in composition and nutritive value, depending on the proportion of high value grain to low value awns and chaff. Malt powder, where it is available separately, may have a relatively high nutritive value but its consistency is that of a very fine dust, and in this form it is very unpleasant to handle. Pelleting removes the physical disadvantage of malt powder and the co-product can then be marketed on its nutritional merits.

Most commonly, malt powder is combined with malt culms (see later) and the mixture is marketed as a feed under the name malt residual pellets.

On its own, unmalted barley is an unsuitable material for making beer. It lacks the necessary enzymes for brewing, the friability for easy milling, and it produces a highly viscous extract that is deficient in amino acids and lacking the required colour and flavour for making beer. Modification of the grain by malting transforms the barley in crucial ways, and the precise malting and subsequent drying conditions determine the type of malt that is produced - the full range of malts stretches across a wide range from pale through amber, to crystal and ultimately black.

In the initial processing, the barley is alternately steeped and aerated over a period of approximately two days in order to hydrate the starchy endosperm and initiate germination. Traditionally, the steeped grain was then cast (spread) onto the malting floor and turned regularly by hand to control the temperature and the rate of germination. More commonly these days, germination is carried out in drums or "Saladin" boxes, which allow regular turning and the consequent aeration of the malting barley to be controlled mechanically. On the floor, germination was allowed to proceed for up to 12 days but, in the boxes, this stage occupies a much shorter period of some four to five days at a temperature of 13-18°C. During this time, a number of enzymes are produced that begin to solubilise the stores of starch and protein within the grain, and to break down the ß-glucans in the cell walls. Breakdown products will be transported to the embryo to sustain further growth and support increased respiration. Inevitably, malting results in a net loss of energy (4-5 per cent by respiration), and the growing embryo also converts some of the sugars (a further 4 per cent) into the cellulose that will form the cell walls of the developing root and shoot. This conversion within the embryo renders some of the sugar unavailable to the brewer, but it will form a digestible part of a fraction that is later used as animal feed.

The principal objectives in malting are the production of enzymes and the breakdown of the cell walls surrounding the starch granules, which render the malt friable. Once these objectives have been achieved, the maltster has no wish to prolong the development of the embryo. Consequently, this latter phase is truncated at the appropriate point by drying, or kilning, what at that stage is known as the "green" malt. In the maltings, germinating barley produces both roots and shoots but the process is normally stopped before many of the shoots protrude through the seed coat (a waterproof protective layer comprising the husk, pericarp and testa). The rootlets are clearly visible at this stage and about 1cm long. After kilning, the dry rootlets are removed and this is most commonly performed in a revolving drum, with perforations that enable the rootlets to be screened off as they break away from the malted grain. This detached, dry fraction is then marketed as **malt culms** to the animal feed industry. Prior to dispatch to the brewer, the malt is usually screened again to remove any thin corns and dust and this fraction, known as **malt screenings**, is also marketed as animal feed usually in combination with the culms.

Kilning is carried out under closely controlled conditions, over a period of 18 to 38 hours. The aim is to stabilise the product with minimal denaturation of the enzymes that will be needed in further processing, and to induce in the malt the required colour and flavour which develop as a consequence of browning (or Maillard) reactions between the sugar and amino acid components. The temperature is increased in stages as the moisture content falls to 3-5 per cent, with careful attention being needed for the lager and pale ale malts which require different temperature/time profiles. A combination of higher temperatures and higher moisture contents is employed when highly coloured malts are required.

The kilning conditions may have implications for the digestibility of the spent grains that will be separated as animal feed after mashing, but there appears to be very little information on this point. In order to explore this question, a small study was carried out specifically for the purpose of writing this chapter. The details given in Table 3 suggest that, in the majority of cases, different malts show only minor variations in digestibility (NCGD value). However, extreme samples - as represented by the roasted barley and black malt – do indicate slightly lower NCGD values. N.B. The various malts were chosen at random as being representative of different types; the results should not be interpreted as the effects of different malting procedures on the same barley sample.

Table 3: The composition and nutritive value of a range of barley and malt samples*

Material	*Protein* %	*Oil* %	*Starch* %	*Sugars* %	*NDF* %	*NCGD* %
Raw barley	10.4	3.3	55.5	3.1	20.6	84.0
Roasted barley	13.5	2.8	57.1	5.0	19.1	81.3
Food grade malt	9.3	3.5	62.1	5.8	17.7	89.7
Pale malt	10.4	3.2	42.2	12.3	18.1	88.8
Lager malt (B)	10.5	3.2	62.2	10.3	15.0	88.6
Crystal malt	10.7	2.4	57.2	13.3	20.4	87.0
Lager malt (M)	10.6	2.9	56.2	9.1	21.3	85.4
Black malt	12.9	2.9	53.6	3.7	21.2	80.9

*All values expressed on a dry matter basis.
The barleys and most malt samples were supplied by Muntons plc; the Lager Malt (B) was provided by Bass Brewers Ltd. The analyses were carried out by NRM Ltd.

Mashing

In traditional practices, the malted barley is first cooled, and it may be rested for some weeks before use. It is then lightly crushed into a coarse meal, referred to by brewers as the grist, and other sources of starch may be added at this stage. These adjuncts may comprise whole cereals such as wheat or unmalted barley, or cereal fractions such as maize grits. The grist is then transferred to the mash tun where it is infused for approximately one hour with water at a steady 65°C or, in lager brewing, the grist is more commonly infused at an increasing temperature of 50 to 65-70°C. During this stage, the gelatinisation of the starch is completed and much of it is converted by mixed enzyme attack into fermentable mono-, di- and trisaccharide units. Proteolysis, which began during malting, is continued during mashing so that, by the completion of this stage, as much as 35-40 percent of the initial barley protein has been converted into polypeptides and free amino acids. Both of these protein breakdown products form part of the liquid phase - known as the wort - and they will transfer with the sugars and other soluble fractions into the next stage of the brewing process. However, the overall effect of the mashing procedure is a net increase in the crude protein content of the spent grains that remain after mashing, because the solubilisation and removal of the substantial carbohydrate fraction is much more extensive than that of the protein.

The effect of mashing on the composition and nutritive value of all of the materials shown in Table 3 was examined using the Institute of Brewing's standard mashing procedure (IOB 1997) in the Bass Technical Centre laboratory. As expected, the protein and oil contents of the grains were substantially higher than the corresponding values in the ingoing materials, although filtration problems confounded the results of this experiment. However, it was notable that the digestibility (NCGD) of the spent grains from the crystal and black malt samples, at 47.0 and 43.8 per cent respectively, were substantially lower than the typical value of other grains. N.B. The average NCGD value of commercial samples of brewers' grains tested by James & Son (unpublished) is 64 per cent. Such reductions in digestibility value give an indication of the effect that extreme kilning conditions may have on the feeding value of the resulting spent grains. Low digestibility has been recorded in very specific commercial situations where roasted malts have been used in the production of dark ales, and the author has found NCGD values as low as 33 per cent in samples of **black (spent) grains**. However, dark-coloured ales are not all produced from roasted malts and, in normal practice, when highly coloured malts are used at all, they comprise only a small proportion of the brewers' grist.

Filtration

With traditional infusion mashing, the sugar-rich wort is drawn off at the end of the process through slots at the base of the mash tun, and the extracted grains act as a thick filter bed. This bed is then sparged with hot water in order to displace the sugar solution held between the particles and to encourage further diffusion of sugars from within the cells. After filtration, the solid fraction plays no further part in the brewing process; it is commonly referred to by brewers as spent grains and is transferred to a silo for use as animal feed. The animal feed industry has long been familiar with this material as a valuable co-product of the brewing process, and typically refers to it by the term **brewers' grains**. This term is also used for the spent grains co-product from industries other than brewing - including malt extraction for a number of non-alcoholic beverages, for malt vinegar manufacture and for the distillation of potable spirit. All of these processes are similar up to the wort production stage and their output of spent grains is also broadly similar.

The last runnings from the filtration vessel may contain a relatively high proportion of lipid and tannins and this fraction, which is not wanted by the brewers, is either diverted to waste or left with the extracted solids. As a potential contributor to the nutritive value of the spent grains, the last runnings may have significant value. There is limited information on the composition of this fraction but one sample was found to contain 34 per cent protein in the dry matter, and a relatively high NCGD value of 74 per cent (James & Son, unpublished information). However, the low dry matter content of last runnings - less than ten per cent - is a negative feature, and the desirability of its inclusion may depend on the ability of the grains to absorb this liquor without subsequently giving rise to effluent problems on farm. That ability will depend in turn on the achievement by the brewery of a satisfactorily high dry matter content in the grains themselves.

Wort separation is a slow process and other methods of filtration have now been introduced. In many breweries, the mash is pumped after extraction of the sugars into a lauter tun where the wort is separated, and rotating knives continually cut the bed of spent grains in order to speed up the process. From the animal feed viewpoint, no significant difference has been recorded between the quality of the grains from either a mash or a lauter tun. More recently, the lauter tun has been replaced in a number of larger breweries by mash filter systems, in which a thin bed of grains and a series of polypropylene cloths provide the filter. For such a filter to work effectively, it has been found necessary to reduce the pore size of the mash bed by hammer milling the malt to produce a fine grist, and to apply additional pressure. Modern mash filters such as the Meura 2001 have been in production since the late 1980's, and they continue to grow in significance. They have brought productivity benefits to the brewhouse, but they have also brought changes to the composition and the physical properties of the extracted solids, which are sometimes identified specifically as **mash filter grains**.

The major difference to the spent grains brought about by the introduction of a mash filtration system – perhaps better defined as a mash press or filter press system – is a significantly higher dry matter content. One stated objective of the Meura 2001 mash filter is the production of a "dry spent cake of at least 30-35 per cent dry solids" (Jones, 1992). In practice, such dry grains have sometimes proved difficult to move within the brewery and mash filter grains have a typical dry matter content of around 27 per cent (D. Clifford, Meura (Brewery Equipment) Ltd; personal communication). However, during the commissioning period for a new system, a much wetter co-product has sometimes been experienced (19-23 per cent dry matter) and this has led to great practical difficulties on farms. Traditionally produced brewers' grains retain a proportion of the sparge water, and this tends to leak out when the grains are stored on-farm. Although this effluent has a high biochemical oxygen demand (B.O.D.) of 30-50,000 mg/litre and is thus a potential pollutant for the river system (MAFF, 1991), it is easily collected and disposed of safely. By contrast, mash filter grains of similar dry matter content present a much greater problem. Instead of the effluent flowing into the silo drain, the whole product tends to move, blocking the drain and spreading out over the farmyard as a thin layer of unfeedable waste. Inevitably, storage losses are high in such situations, and the problem is made worse by aerobic decay, since this occurs more rapidly in the finer material which has a greater surface area exposed to the air. The instability of a stack of such low dry matter grains makes it difficult to seal, and this compounds the aerobic spoilage problem. Such practical difficulties provide a timely reminder that changes within a brewery should not be made without due regard for all the implications.

Variation in spent grains

Brewers' grains from different sources do vary in composition and nutritive value, although supplies may be more consistent within breweries. Hall (1995) gave protein and energy values for brewers' grains from nine sources, in which average protein levels ranged from 19 to 30 per cent and the estimated ME values for ruminants varied from 11.4 to 12.3 MJ/kg on a dry matter basis. In vivo studies in Aberdeen with six samples of brewers' grains found an even wider spread of ME values; from 10.9 to 12.5 MJ/kg DM (Rowett Research Institute, 1984). The differences can be attributed to the range of raw materials that is used and associated differences in extraction efficiency. Minor variations in brewers' grains composition may be caused by differences in the handling of last runnings and trub (see later). The introduction of a mash filter system, with an improved efficiency of extraction, may be expected to reduce the quality of the spent grains but no substantial differences have been noticed in practice – the significant effect on dry matter content has already been noted.

The tables of composition and nutritive value that form part of this book include separate data for brewers', mash filter, vinegar and malt extract grains. To a large extent, and for many of the parameters, the data show ranges with a considerable degree of overlapping. However, the protein content of the two latter types appear to be lower and the starch content higher than spent grains from the brewery. These differences in composition may reflect differences in both the selected raw materials and the efficiency of extraction.

A deliberate reduction in the moisture content of brewers' grains was advocated by Penrose in 1982 – a practice that is usually referred to as dewatering, though the separated liquor always contains nutrients and potential pollutants. He noted that screw presses had been used for many years in American and European breweries but, for large breweries, Penrose suggested that centrifuging by means of a decanter would be preferable. For him, the motivating factor was a potential improvement in the brewers' extract, not the production of a higher value feed material, and the absence of such systems in UK breweries today, suggests that the value of the improved extract did not compensate for the necessary investment in this process. However, the screw-press would potentially represent a relatively modest expense and its use in UK breweries may need to be reconsidered, if only to reduce the high cost of hauling unnecessary amounts of water from brewery to farm. An example of such a press is in use at the Novartis malt extract factory, where this simple processing results in an increase in the dry matter content of the grains by approximately two percentage units.

For the brewer, the use of a screw press may result in the inclusion of undesirable substances in the centrate (Penrose, 1982), making the extract unsuitable for his purposes. Consequently, the press would need to be positioned at a later stage and, in this situation, the cost of the additional processing may have to be justified by increased returns from the feed market. However, since pressing would result in a reduced quantity of grains carried over the weighbridge, the drier product would need to command a higher price even before the cost of the operation was calculated, and the disposal of the pressed juice may represent an additional expense. Such considerations possibly explain why this potentially attractive development for the animal feed market has not received more attention. Looking to the future, it seems possible that environmental concerns associated with the possible pollution threat of high moisture grains, together with the increasing cost of haulage, may necessitate a re-consideration of dewatering practices.

The choice of pressing equipment would depend on a number of factors but a balance may need to be struck between the nutritive value of the solid and liquid fractions. A positive feature of the Novartis processing is that, while the grains find a ready demand for a drier solid, the liquid has a sufficiently high value to be considered as a feed not an effluent. The expressed juice is currently being marketed as a high protein liquid feed for pigs, under the name "**grains pressings**". The composition of the press liquor depends *inter alia* on the size of the holes through which the juice is pressed - smaller holes are claimed to yield lower dry matter juice (R Wurm, Vetter Maschinenfabrik; personal communication). At Novartis, where a cylinder press forces the liquor through holes of 3mm diameter, it has a typical dry matter content of 9 per cent. If the dry matter content of the juice were lower, the liquid co-product may be more difficult to market as a feed.

Adjuncts

The amylolytic capacity of most malts is more than is needed to convert the malt starch into fermentable sugars. This spare enzyme capacity permits the addition of other sources of starch, which are normally less expensive and can bring additional benefits in terms of beer quality. Almost any starch source can be used for this purpose, although there are legal restrictions in some countries. In Britain, the choice will be determined by a number of factors including cost, extract potential and typical extract efficiency, plus the content of undesirable lipid and the effects of the adjunct on wort clarity, viscosity and filtration rate. Most cereal grains will need a degree of preliminary processing to ensure that the starch is accessible to the enzymes. Barley and wheat flour can be used directly because the mashing temperature is sufficient to gelatinise their starch, but rice and maize need to be cooked prior to use. However, despite tailoring the processing conditions to the specific needs of the adjunct, spent grains always contain a proportion of ungelatinised starch (Lloyd, 1986).

Throughout the last century, starch-rich adjuncts have been quantitatively less significant to UK brewers than those based on sugars (J. MacDonald, Bass Brewing; private communication). Although a small proportion of adjunct sugars, such as diastatic malt syrup, may be added to the mash tun, syrups are most commonly used as wort extenders and they are included only after the separation of the spent grains. Some priming sugars and caramels are even added after the main fermentation, in order to adjust the colour and flavour of the beer and to permit a degree of cask-conditioning. These sugar-rich adjuncts are consequently of little significance to the animal feed industry, where the greatest interest lies in the proportion of the adjunct that is not solubilised and extracted but remains after mashing and filtration as part of the spent grains fraction.

An indication of the value to the animal feed industry of a number of brewing adjuncts is given in Table 5, extrapolated from the information provided by Lloyd (1986).

Table 5: Residual value of various brewing adjuncts*

Grist ingredient	*% Extract*#	*% In spent grains*
Raw barley	64.7	35.3
Torrefied barley	69-72	28-31
Flaked barley	72.0	28.0
Raw wheat	76.4	23.6
Torrefied wheat	77-78	21-22
Flaked wheat	78.2	21.8
Ale malt	80.3	19.7
Lager malt	81.0	19.0
Maize grits	88.6	11.4
Wheat flour	88.6	11.4
Flaked maize	89.2	10.8
Rice grits	92.0	8.0
Flaked rice	92.5	7.5
Refined starches	102-105§	0.0

* All values refer to the dry matter.
# Some values have been converted from expression in L°/kg by the standard equation: % Extract = (0.2601 x L°/kg) – 0.3064
§ Values in excess of 100 per cent represent chemical gain during the hydrolysis of starch to sugars.

While the brewer concentrates attention on the extract figures, the residual amounts shown in the third column of Table 5 are more relevant to the brewers' grains merchant and to the farmer who purchases livestock feeds. Thus the addition of unmalted barley can be expected to yield a significant fraction for animal feeding purposes, whereas the use of rice adjuncts leaves a much smaller proportion, and the use of refined starches or sugar syrups leaves nothing other than malt residues for the feed market.

The contribution of adjuncts to brewers' extracts varies from country to country, brewery to brewery and between individual beers in a single brewery – adjuncts have contributed as much as 40 per cent of the fermentable solids in the USA, but only about half that amount in the UK (Lloyd, 1986). Thus the impact of adjuncts on the animal feed market depends on two factors: the amount they contribute to the grist and the proportion that remains after wort filtration.

Hops and trub

After separation of the spent grains, additional fermentable sugar in the form of a syrup is often incorporated into the wort, and hops are added to induce the desired flavour and aroma in the beer. The hop plant, Humulus lupulus, is dioecious – bearing both male and female flowers on separate plants – but it is the female inflorescence that is used in brewing. In these cone-like structures, the sticky yellow lupulin glands contain resins and essential oils that will impart the required characteristics. Hops marketed within the European Union are subject to statutory certification standards, which include maximum limits on any contaminants. Pesticide residue limits (MRL's) also apply to hops, as they do to the cereals used in brewing, and individual breweries carry out random checks to ensure compliance. Chemical tests are supplemented by visual checks, which ensure that no contamination has occurred en route to the brewery.

Whole hops now make up only 10 per cent of the hop supply (BLRA 2000), much more popular are hop pellets (60 per cent) and hop extracts (25 per cent). Compared to the unprocessed hop, the greater efficiency with which these hop products transfer their characteristics to the wort implies that a lesser proportion will remain unused; i.e. a smaller amount will be left as a co-product fraction.

The supplemented wort is then boiled to ensure sterilisation, the inhibition of further enzyme activity and the development of the required bitter flavour. Boiling also serves to coagulate some of the proteins that originated in the grist or the hops, though the retention of some amino acids in solution is essential for yeast growth in the fermentation stage. The fine sediment in the boiled wort must be removed to avoid the beer becoming cloudy, and this fraction, which also contains unwanted tannins, resins and minerals from the hops, is typically flocculated. A flocculating aid derived from an edible seaweed (Chondrus crispus) is often added towards the end of boiling. The active principle of this seaweed is an acidic polysaccharide, carrageenan, which is used in food as a permitted emulsifier (E407). In brewing, a semi-purified, powdered form of the material - of defined particle size and chemical quality - is used at 40-80 mg per litre, and this interacts with coagulated protein to encourage flocculation. The floc, together with the flocculating aid, settles out to form a residue commonly known as the trub, and this process leaves a clear wort to go forward for fermentation. Trub has undoubted feed potential – see Table 6 - though its protein content appears not to be highly digestible. The composition may vary in relation to the form in which hop products are used. In particular, where whole hops are used, the fibre content of the trub can be expected to be higher.

Table 6: Composition of trub

Crude protein (% DM)	35.4
Protein digestibility (%)	61.9
Crude fibre (%DM)	2.3
Oil (%DM)	1.5
Ash (%DM)	7.4

Source: Malting and Brewing Science, Volume 2.

However, it is difficult to pump trub from the "whirlpool" separation vessel without slurrification, and this process reduces its feeding value in direct proportion to the volume of water added. In the past, trub has either been disposed of as a liquid waste, or it has been added to the brewers' grains. Trub could enhance the nutritive value of the grains, although its bitter taste may be unpalatable. Its incorporation into brewers' grains could lead to effluent problems on-farm unless it is metered in carefully with due regard to the volume of water that has been used.

Fermentation

The clear wort is cooled and transferred to the fermentation vessel. Traditionally these vessels were relatively small, and constructed from a range of materials but modern breweries employ much larger fermentation vessels, and because of cleaning requirements, they typically rely on stainless steel. However, despite the increase in scale, the fermentation in most breweries remains a batch process - only a minor proportion have adopted a continuous system. The wort is inoculated with a specific culture or a mixed culture of yeasts, containing two or more strains of *Saccharomyces cerevisiae* or *Saccharomyces carlsbergensis*. Yeasts are selected for their ability to produce the characteristic aroma and flavour in the beer although, essentially, their prime purpose is the conversion of the wort sugars into alcohol and carbon dioxide. Most brewers have developed their yeast strains in-house, and have maintained them over a prolonged period, stretching back perhaps as much as two hundred years (J. MacDonald, Bass Brewing; private communication). At approximately 10-15 million yeast cells per millilitre, the inoculum is equivalent to the addition of 3 kg pressed yeast per thousand litres of wort.

Fermentation does not occur immediately but must await developments within the yeast cell, wherein both oxygen and amino acids are absorbed from the wort and enzymes are synthesised. Cell multiplication normally begins after a lag-phase of 12-20 hours, and the yeast commences to assimilate sugars – firstly the monosaccharides, followed by maltose and sucrose, and lastly the trisaccharides. No sugars of greater complexity than maltotriose can be assimilated, which confirms the importance of the mashing procedure. Over the next two to three days the yeast population will increase five-fold, and this proliferation will be accompanied by vigorous fermentation. The yeast is repeatedly "cropped" during fermentation in order to reduce the likelihood of inactive yeast cells dying and possibly tainting the beer. Brewers are conscious of variation in the quality of the yeast, and they typically select the middle "crop" with which to inoculate the next fermentation. Eventually, the rate of cell multiplication declines and, although fermentation continues, the yeast cells begin to lose weight before ceasing their activity completely.

Yeast is separated from fermenting wort either by flotation or sedimentation, depending on the flocculating properties of the yeast and the type of beer being produced. Once separated, the yeast is normally pressed to remove the beer and, at this stage, it has a typical dry matter content of up to 30 per cent. Approximately 15 kg of pressed yeast will be produced per thousand litres of beer, and perhaps 20 per cent will be re-used within the brewery. The surplus will be marketed as **brewers' yeast**, either to the food and pharmaceutical industries or to the animal feed market. It is sometimes found necessary to slurrify the pressed yeast with additional water to enable the material to be pumped out, but slurrification inevitably reduces the feeding value of this co-product in proportion to the volume of water added. The brewers' yeast typically delivered to the farm has a dry matter content in the region of 11-16 per cent.

The viability of the yeast is a crucial requirement of the brewing system, but the yeast that is surplus to the brewer's requirements must be rapidly killed if it is not to cause problems subsequently. Given a supply of suitable carbohydrate, live yeast continues to produce carbon dioxide and this leads to profuse frothing, with the material bubbling out of the smallest orifices to create an environmental nuisance. Severe animal problems, including pig deaths, have also occurred where live yeast has been fed in conjunction with a feed containing a significant amount of sugar, as a result of fermentation within the gastro-intestinal tract (A J Walker, 1983).

Thus it is common practice for brewers' yeast to be killed at the brewery, usually by the application of formic and propionic acids. But there can be a significant cost to such treatment, and occasional supplies of live or inadequately treated yeast do appear on farms. As an animal feed, brewers' yeast is highly prized as a quality protein source, with protein and lysine contents on a par with soyabean meal. Its availability for animal feeding depends on the demands of the human food and alternative markets, but the potential yield is 24 kg per thousand litres of beer (at 15 per cent dry matter content), in addition to that re-used within the brewery. For an annual volume of 5.66 billion litres of beer one can thus calculate a theoretical yield of 135,000 tonnes of brewers' yeast – equivalent to 68,000 tonnes in the pressed form.

Beer is just one example of the brewers' art
(courtesy: Novartis Consumer Health)

Beer

The end-product of brewing is a range of different beers, with lager in recent years becoming increasingly predominant (BLRA 2000). However, almost two thirds of beer is marketed in draught form and lager and ale/stout make up roughly equal proportions of this sector. Some of this cask, keg and tank beer is rejected for a number of reasons, usually related to its appearance and taste, and this is returned to the brewery. There the beer may be mixed with some of the output from the yeast press, which is considered to be surplus or out-of-spec in relation to its re-use within the brewery. The reject product then becomes available as a feed, and is usually referred to as **ullage**. Pigs typically find ullage to be a palatable feed ingredient, but the low dry matter content makes it uneconomical to transport over long distances. However, the dry matter content needs to be determined carefully, since conventional oven-drying would result in the loss of alcohol in the oven, and an underestimate of the true dry matter content. Ullage is predominantly an energy feed and the alcohol content gives it a significant energy value.

The co-products of malting and brewing

Barley screenings

Barley screenings represent the thin and broken barley grain, together with a proportion of barley awns and chaff, that the maltster has screened out of the material he selects for malting. To reach this point, barley screenings have had to satisfy a number of checks, including examination for the presence of insects, weed seeds, and moulds as well as soil and other inorganic components. In common with the barley used for malting, barley screenings have also satisfied the maltster's requirements with respect to appropriate chemical treatment in both field and store, and the co-product has been separated from a consignment that carried a grain passport detailing the history of the crop since harvest.

For the feed industry this material can thus be regarded as clean and wholesome, though of variable nutritional value depending on the inclusion rate of awns and other fibrous parts of the barley plant. Variability is also increased by practices that differ between factories with respect to the inclusion of malt dusts that arise in other parts of the maltings. Anyone offered a supply of barley screenings would be well advised to assess the nutritional quality of the material from the particular source.

For cattle, the presence of whole grains is likely to reduce the digestibility of the feed compared with the value indicated by laboratory tests. This is due to the passage of whole grain through the digestive tract in a largely undigested form. Grinding the material before feeding would largely remove this problem, but it would also increase the dustiness of an already dusty co-product. Pelleting improves the handling characteristics of barley screenings and increases their bulk density - with additional benefits in terms of reduced haulage costs.

The proportion of non-grain material strongly influences the fibre content of barley screenings and has an inverse relationship with the energy value of the feed. The protein content may be expected to be relatively low and poor in lysine. For these reasons, this co-product is best regarded as a feed for ruminant animals. Sheep could potentially make better use of whole grain than cattle; in fact they have been shown to use whole grain as efficiently as the processed form (Morgan *et al.*, 1991). Thus barley screenings, unless ground and pelleted, may be a better feed for sheep than cattle.

Malt powder

Malt powder is collected from various points as the malt is moved through the conveying system and other parts of the malting plant. It has a very variable composition, but is typically rich in husk and husk components. The dry matter content is invariably high and values of 88 to 96 per cent have been recorded (James & Son, unpublished data). Compared to barley grain, malt powder tends to have more protein and oil, which add to its nutritive value, but also more fibre and ash, which are a significant detraction. Digestibility (NCGD) values vary widely; from 56 to 90 per cent (James & Son, unpublished).

It is the physical nature of malt powder, rather than its chemical composition, that is the principal feature of the co-product. Malt powder is typically a fine, dusty material with a low bulk density. It can be expensive to transport, unpleasant to handle and wasteful to feed. In order to overcome these difficulties, the co-product is often pelleted – commonly in combination with the malt culms that are produced on the same site. The resulting mixture is marketed as malt residual pellets.

Where malt powder is to be handled in loose form, it can usefully be included in a moist mix, with silage, liquid potato feed or a distillery syrup. In the author's experience, malt powder has been included at 15 per cent of the dry matter in a moist concentrate blend, where it has provided a useful source of nutrients in a satisfactory physical form.

Malt culms

Malt culms are the dried rootlets of the germinating grain, whose development is initially encouraged and then arrested by the maltster, once the malt has been adequately modified and has developed the required enzyme capability. The rootlets are killed by drying and detached from the malt to become available as a co-product animal feed. The drying procedure results in the culms having a shrivelled appearance, but a very high dry matter content - in the region of 93 to 96 per cent.

This feature of the co-product – ten per cent more dry matter than the original grain – implies an increase in the nutritive value by a similar amount.

However, malt culms have a very different composition from the parent material. They have a low starch content and much more fibre than either the original barley or the malt from which they have been separated. But malt culms are much richer in protein than either of these two raw materials, and it is principally the protein content that establishes their value in the animal feed market. In common with other cereal derived feeds, protein quality is not high, and malt culms are more suitable as a protein source for ruminants than for non-ruminant stock. In terms of the supply of essential amino acids, malt culms are in fact poorer than barley, when considered in relation to the feed's higher level of crude protein. Although the lysine content of the protein is in line with that of barley, the methionine, cystine, threonine and isoleucine contents are all poorer.

Malt culms have a pleasant aroma but some are reported to have a bitter taste (Lonsdale, 1989). This is not a universal finding, though the unpalatability of some peaty malt culms is well recognised. Most other sources of malt culms would be readily accepted when fed as part of a typical mixed diet. This feed also has a very low bulk density, which can translate into a high haulage cost. Principally for this reason, and to reduce dustiness, malt culms are commonly mixed with malt powder and pelleted before the feed leaves the maltings. This blend of co-products is marketed under the name Malt Residual Pellets, though neither component can truly be regarded as a residue.

Malt residual pellets

Malt residual pellets (MRP) are a blend of two co-products from the malting industry, malt powder and malt culms. In their natural state, both these co-products have a low bulk density, and this implies a high haulage charge for the delivery of such lightweight materials. Malt powder is also a difficult and unpleasant material to handle and this disadvantage tends to obscure its nutritional value. Pelleting is a recognised means of increasing density and it transforms these two feed materials from "difficult" into "useful".

Nutritionally, the two components complement one another well; the malt powder is typically a source of starch, while the culms are essentially a protein supplement. Together, MRP becomes a medium quality material with a variety of uses in ruminant diets. Mixed in equal parts with cereals or sugar beet pulp, or the equivalent in moist potato or citrus pulp, MRP could provide a substitute for part of the compound feed in diets for beef and dairy cows. For young stock reared on straw, the protein content of MRP would make it a valuable supplement, whilst for sheep MRP could represent a safe, relatively low-starch alternative feed for animals unaccustomed to a concentrate diet.

Brewers' grains

Brewers' grains are the solid residue left after the extraction of malt in the production of beer, malt extracts and malt vinegar. Other starch sources, such as maize, rice, wheat and barley are sometimes used as adjuncts to the malt and a proportion of these materials will also be found in the grains. After filtration, little of the original cereal starch remains in the grains, but the protein, fat, fibre and mineral components are all concentrated in this solid fraction. The gross energy value of brewers' grains is high - a range of 20.7-21.9MJ per kg DM is given in MAFF Tables (1990) from measurements made on 20 samples – and this is largely a reflection of the high oil and protein contents of grains. Digestibility is relatively low – in vivo DOMD values of 54-63 per cent were found in a study of 6 different consignments - and this is consistent with a feed that comprises 50-75 per cent neutral detergent fibre and 14-19 per cent crude fibre.

Brewers' grains are a traditional feed
(courtesy: James & Son (Grain Merchants) Ltd)

Brewers' grains are a palatable feed that has been used for centuries as a succulent ingredient of rations for cattle and sheep. The energy value contributed by the grains is largely in the form of oil and digestible fibre, and research at the Rowett Research Institute in Scotland has shown how efficiently this feed material is digested in the rumen (Rowett, 1984b). There is a much lower wastage of energy as methane gas than occurs when the whole of the grain is fed (c.f methane losses of 4 v 10 per cent, MAFF, 1990). Further energy efficiency may be expected at tissue level when the animal utilises ME derived from oil in place of alternative carbohydrate sources. This greater efficiency may explain why some farmers have reported a lift in animal performance when brewers' grains were introduced into the ration.

Owing largely to the range of raw materials used by different brewers, the composition of brewers' grains varies between different sources, but tends to be more consistent from any individual supplier. In the author's experience, crude protein levels have ranged from 17 to 32 per cent, around a mean value of 24 per cent of the dry matter. The ME values measured in an in vivo study with sheep at the Rowett Research Institute (Rowett, 1984b) ranged from 10.9 to 12.5, with a mean of 11.7 MJ per kg dry matter. It is notable that the ME value recommended by independent advisers has increased as evaluation procedures have improved; Technical Bulletin 33 (MAFF, 1976), which was the advisers' guide at that time, quoted an ME value of only 10 MJ per kg dry matter. N.B. The ME value of brewers' grains may be predicted from equations based on NDF, ADF or NCD, which were developed in the Rowett study. However, the ME value may be significantly underpredicted by equations developed for more general use such as the E3 or Compound Feed equation (Thomas *et al.*, 1989; MAFF, 1993).

As a mineral source, brewers' grains are notable for their phosphorus content, and the feed can be a valuable source of this important and expensive mineral. Specially formulated mineral supplements may be needed in order to maximise this benefit when large quantities of brewers' grains are fed.

Brewers' grains can be fed fresh from the day they arrive, or ensiled for feeding some months later - in both cases the feed is attractive to stock. Even with short-term storage it is important to store the grains carefully, to reduce the risk of aerobic spoilage. The freshly delivered co-product contains a proportion of free water – added in the brewery or factory to extract the remaining sugars from the grain – and some of this drains from the grains when they are allowed to stand. Although this colourless liquid may contain only two per cent dry matter, it has a B.O.D. level in the region of 30-50,000 mg per litre and is a potential pollutant unless collected and disposed of safely (MAFF, 1991 and 1998). Alternative storage procedures have been developed that enable much of this juice and its accompanying nutrients to be retained, such as the incorporation of 13-15 per cent of dried sugar beet pulp with the grains at ensiling. The resulting mix is referred to as Grainbeet, and this farm-blended feed has been used successfully for all classes of ruminant livestock (Trident Feeds).

Brewers' grains can be fed fresh or ensiled for later use
(courtesy: James & Son (Grain Merchants) Ltd)

Principally, brewers' grains are used as a protein source that boosts the value of rations based on maize silage, straw and other low protein roughages. They are also used strategically by dairy farmers to manipulate milk composition. Research has demonstrated that brewers' grains, whilst stimulating both milk yield and milk protein, can reduce the milk fat content (CEDAR, 1995), and the production of a greater quantity of lower fat milk can represent a profitable strategy. Dairy cows are typically fed 9-10 kg brewers' grains, although much higher feeding levels are possible. In an internal MAFF report, probably dating from the 1950's, Brown and Eden described a system of self-feeding brewers' grains in which dairy cows consumed an average of 23 kg of stored grains – equivalent to approximately 6.4 kg brewers' grains dry matter per head per day. The CEDAR trial referred to above included grains in place of one-third of the silage on offer, and cows ate approximately 17 kg brewers' grains per head per day.

The CEDAR trial also highlighted another feeding option, in which brewers' grains are used in place of a proportion of the basal grass silage ration. This may offer a viable, economic alternative to a total reliance on silage feeding which, in some situations, is becoming increasingly expensive. An extreme, zero-silage, version of this option was put into practice as long ago as 1968, when a Buckinghamshire dairy herd was fed brewers' grains, sugar beet pulp and wheat straw as the basal ration (T Bucknell, Manor Farm, Haddenham; personal communication). Brewers' grains have also been used in conjunction with silage as a buffer feed to supplement summer grazing, and this blend proved superior to a supplement consisting of silage only (K Aston, Grassland Research Institute; personal communication). Offering an equal part mixture of silage and brewers' grains enabled stocking rates to be doubled without any reduction in milk output.

Large amounts of brewers' grains may also be fed to beef cattle. In a study at Gleadthorpe Experimental Husbandry Farm (ADAS, 1970), yearling Friesian steers on an intensive ration of mineralised barley were given brewers' grains in place of 20, 30, 40 or 50 per cent of the barley. Animal performance was maintained at 1.3-1.4 kg per day on diets containing 0-30 per cent grains, though the growth rate was progressively reduced with higher inclusion rates. Lewis and Lowman (1989) fed draff (a distillery co-product of similar type to brewers' grains) as the sole diet to finishing cattle but recorded a mean growth rate of less than 0.9 kg per day, from an intake which averaged two percent of liveweight in dry matter terms. Although the cattle in this particular study failed to respond to a supplement of 1 kg of barley or sugar beet pulp, offered over the last month of the trial, it would seem fair to conclude that draff (and probably brewers' grains), when fed alone, does not provide the optimum diet for cattle required to finish at a high rate of gain. However, when stored as Grainbeet, and offered as the sole diet to fattening cattle, bull beef animals at SAC Crichton Royal grew at 1.6 kg per day and outperformed another group fed 3 kg concentrate feed with silage to appetite (Trident Feeds). Experience by James & Son representatives in the field suggests a useful synergy between brewers' grains and potato co-products, giving a number of advantages in terms of co-storage, palatability and feed intake.

Sheep can be fussy eaters, but they thrive on brewers' grains at all stages - in pregnancy, lactation and as weaned lambs. Even before weaning, lambs accompany their mothers to the feed trough and have been observed to consume brewers' grains from two weeks of age (J Spencer, Hill Farm, Daventry; personal communication). The fibrous nature of the feed adds an element of safety for animals that are unaccustomed to handfeeding. When offered in the form of Grainbeet, it may be necessary to restrict the supply during mid-pregnancy to avoid ewes becoming over-fat. But for finishing lambs Grainbeet can be offered ad libitum; SAC Auchincruive recorded growth rates of 174 g per day in Blackface stores offered Grainbeet to appetite (Trident Feeds).

Grains pressings

Grains pressings are a liquid feed currently available only from the Novartis Malt Extract factory but, in principle, such a material could be produced more widely. The co-product is produced by the screw-pressing of brewers' grains, and this action separates a high protein juice from a more fibrous solid material. The composition of the juice may be significantly affected by the size of the aperture through which the juice is expelled; at Novartis, the juice has approximately nine per cent dry matter and that dry matter has a fibre content (NDF) of only 8 per cent.

Grains pressings are a nutritious and highly digestible feed, suitable for all classes of stock but perhaps most suitable for pigs fed on a liquid feeding system. The protein level is typically 35-40 per cent of the dry matter, but the amino acid composition may be expected to reflect the co-product's cereal origins, in which case the lysine content would be much lower than that found in protein sources such as soya and rapeseed meals. However grains pressings, with an oil content of 7-8 per cent and a digestibility (NCGD value) of 92 per cent, have an energy value that is substantially higher than either of those alternative protein-rich feeds.

Brewers' yeast

The co-product brewers' yeast available to livestock farmers is identical to that re-used within the brewery to ferment sugars into alcohol – although it has preferably been treated to prevent further fermentation. In the brewing process, yeasts grow and multiply with the result that the brewer produces some five times as much as he/she needs for the subsequent brew. This surplus yeast has traditionally been used as a highly flavoured ingredient by the food industry, but variable quantities of the surplus have long been available for animal feeding, and the co-product has been used for both pigs and ruminants. The yeast is normally available in liquid form, with a dry matter content of the order of 12-16 per cent, but pressed yeast may sometimes be available with a dry matter content of up to 30 per cent. The liquid quickly separates into two distinct layers, and should preferably be stirred immediately prior to its incorporation in a ration. The unstirred solids can pose a challenge to pumping systems on both the delivery vehicle and farm tank.

Live, untreated yeast can pose different problems since it continues to ferment substrates as long as they are available. The carbon dioxide liberated by fermentation causes considerable frothing, and this leads to an environmental nuisance as well as carbohydrate loss, a reduction in true protein content (Steckley *et al.,* 1979) and potentially severe gastro-intestinal problems in pigs fed the yeast in this form (Neame, 1981). In most breweries, the yeast is treated with a mixture of organic acids to kill the yeast cells prior to dispatch, although this adds substantially to its cost. However, when treated in this way, the co-product may then be valued on its merits as a high protein feed, with an unusually high phosphorus content and a rich supply of B vitamins.

Liquid feeds are usually more suitable for feeding to pigs than to ruminants, and the high protein quality of brewers' yeast makes this doubly so. The lysine content - equivalent to 6.6 per cent of the crude protein - is similar to that of soyabean meal. Neame (1981) reported that, when first introduced, the pigs take a few days to accustom themselves to the feed but they subsequently develop a considerable taste for it. Neame's pigs were restricted to a maximum of 0.43 kg yeast dry matter per day, but this amount represented approximately one-third of the diet when this limit was first reached. It was noted that yeast-fed pigs become soporific, and this may represent an additional benefit in reducing the stress associated with a group environment.

Brewers' yeast has also been fed to dairy and beef cattle. Dawkins and Meadowcroft (1962) reported early work in Germany and Russia in which live yeast was fed to dairy cows for three days with an immediate and lasting improvement in the butterfat content of the milk. However, in British trials, these workers fed 0.45 kg per day for 28 days or 10 kg over a 6-day period but failed to show any benefit in milk yield or quality. In a more conventional feeding trial, Grieve and Burgess (1977) compared brewers' yeast with soyabean meal as a supplement to maize silage, for growing steers. A substantial response was associated with both supplements, the growth rate improving in a 98-day trial period from 0.85 kg per day on the control diet to 1.33-1.36 kg per day. The amount of yeast fed in this trial was 5 litres per day and provided 9 per cent of the dietary dry matter. For dairy cows, Roth-Maier (1979) suggested that up to 15-20 litres of "liquid yeast" could be fed in place of soyabean meal and a proportion of the basal ration.

Ullage and other beer

Ullage is the term usually given to feed quality beer, much of it draught product that has been returned to the brewery because of perceived problems of appearance and flavour. At the brewery, it may be mixed with other sources of out-of-spec product, though care must be taken to ensure that this does not include any cask washings that may be contaminated with cleaning fluids.

Ullage is a low dry matter feed with an alcohol level that may vary from 3.5 to 8 per cent by volume. Since ethanol has a DE value for pigs of 29.8 MJ/kg, the alcoholic strength of the beer is a crucial determinant of the energy value of ullage. An Institute of Brewing paper (Martin, 1982) recommended a procedure for the calculation of the calorific value of beer, but suggested that a rapid value could be obtained by a simple conversion of the gravity value. (Viz: Calories per 100 ml = gravity x 0.84 or 0.89 for highly attenuated (sugar-free) or conventionally fermented beers respectively.) For beer with an alcohol level of 3.5-8 per cent by volume, this equates to a gross energy value in the range 19-26 MJ/kg DM, and much of this may be expected to have a digestibility of 100 per cent.

Ullage containing 4 per cent alcohol has been fed to fattening pigs in an ADAS monitored farm study (A J Walker, 1983) at 4.5 litres per day. Growth rates exceeded expectations and, after a soporific initial period, the pigs' behaviour was unremarkable. In a parallel study on another farm (Anon, 1985) the pigs were said to "spend long periods asleep after every meal". Thus the co-product was believed to reduce the pigs' energy needs for maintenance and activity. However, the inclusion rate needs to be controlled if an optimal contribution to the diet is to be achieved. That this has not always occurred in the past was evident from a report in the Veterinary Record (Peace, 1981), where blood samples indicated an alcohol level of 295-347 mg per cent – some four times the human drink-driving limit! Unfortunately, no quantitative information was available on feed intake.

It is possible that excessive alcohol consumption may occur by accident. Ullage may contain a small number of live yeast and, if mixed on the farm with a suitable substrate, fermentation will result in alcoholic enrichment. This possibility should be avoided, because fermentation will also result in the production of carbon dioxide, and this will lead to frothing and increased pressure. In the storage vessel this would be a nuisance, but in the animal it may cause severe colic and even death. It would thus be advisable to avoid using ullage in conjunction with feeds containing a significant proportion of sugar. In particular, ullage should not be stored for subsequent feeding after mixing with any sugar-rich feed.

In recent years, a new source of beer has appeared in the South of England. Some of it was brewed in continental Europe, but much of it was brewed in the UK, exported and then returned. All of it arrives in the feed chain as a consequence of confiscation by HM Customs and Excise. At the point of confiscation, the beer was considered to be of suitable quality for human consumption but, after separation from the drink cans, it is moved quickly as a bulk commodity into the feed chain. Much of this beer supply is of lager type and it will primarily be used as an energy source, with an energy value in line with its alcohol content. However, some pig farmers have recognised that the non-alcoholic fraction of beer also has a monetary value. On large holdings, the water supply can represent a significant cost when metered by the local water authority. Dilute liquid feeds delivered in 24,000 litre tankers can make a contribution to the reduction of this cost.

Malt vinegar production

Vinegar originated as a by-product of alcohol fermentation and different vinegars reflect the types of alcoholic beverages that predominate in different regions. Thus wine-growing countries use wine vinegar, cider makers produce cider vinegar and beer-drinking areas tend to favour the production of malt vinegar. Although the principal product of all these activities may be similar, the co-products of vinegar production have little in common. Thus vinegar co-products need to be considered separately, as defined by their raw material.

Malt vinegar manufacture has evolved in line with the changes seen in beer production. Prior to the 1880's, malt vinegar was the natural by-product of on-farm beer making, and the industrialisation of both processes occurred at the same time. As its starting point, the malt vinegar industry continues to use the inputs that have been developed for the brewing and distilling industries and, even to this day, there is no record of barley being grown specifically for the production of malt vinegar (J Rostron, Nestlé; personal communication). However, the specification for the malted barley used in malt vinegar production is specific for the purpose – although it does bear great similarity to distilling malt.

Reliance is placed on the malt industry to ensure good farming practice with respect to pesticide use and the avoidance of mycotoxins and other contaminants in the raw material. But vinegar producers do require evidence of regular monitoring by all suppliers to demonstrate that acceptably low levels of any potential contaminant are being achieved. The vinegar producers then carry out a degree of cross-checking using samples from loads that arrive at the factory, and it is notable that tests made by Nestlé have never found pesticide levels above the maximum residue limit (J Rostron, Nestlé; personal communication).

In the factory, the selected malt is soaked in hot water in the mash tun, and maintained at a temperature of 60-65°C for an hour in a parallel process to the infusion mashing that occurs in the brewery. At times in the past, other sources of starch - principally other cereals - have been used as adjuncts to the malt but this is not current practice. The malt starch is rapidly gelatinised and then converted by the malt enzymes into a range of simple sugars. A proportion of the malt protein is also solubilised and this will later be a vital ingredient for yeast growth. At the end of the process the sweet wort is drained to leave the extracted grains, and this moist co-product is marketed to the animal feed market.

Animal nutritionists typically draw no distinction between these **vinegar grains** and the spent grains from a brewery. However, the choice of malt and the absence of adjunct starch sources would appear to make them more similar in derivation to the grains from a malt distillery (draff), which are typically considered to be of lower feeding value than brewers' grains. The results of analyses of vinegar grains from the Nestlé factory would not support the implication of lower quality; their composition, in respect of both energy and protein values, appears to fall within the range of values found for grains obtained from breweries.

In the factory after its separation from spent grains, the sweet wort is transferred to a stainless steel fermentation vessel and inoculated with yeast. At this point, the vinegar producer varies from the practice adopted by most brewers. The specific strain of Saccharomyces cerevisiae that is used is entirely cultured in the laboratory – no yeasts are cropped from the previous batch and re-cycled. In contrast to fermentation in the brewery, which proceeds over a period of perhaps three days, the vinegar producer aims single-mindedly for maximum alcohol yield, and maintains his fermentation for five to six days. At the end of this prolonged process, in which the alcohol concentration reaches 9 per cent, the spent yeast is no longer viable. The lack of suitability for re-use means that all of the yeast becomes available for alternative purposes, and it is currently being marketed as a feed for pigs. Since the volume of this co-product is relatively small, the yeast at the Nestlé factory is combined with another co-product that is produced later and it is the blended material that is marketed as a feed.

After the fermentation stage, the yeast is removed and the fermented wort is subjected to a second fermentation, in which the alcohol is oxidised in a continuous process to acetic acid. The acetic acid bacteria used in this process are specific cultures of acetobacter, and the strains are claimed to be "jealously guarded by individual manufacturers" (J. Rostron Nestlé; personal communication). The resulting product of this second fermentation is malt vinegar, a fairly complex cocktail comprising various alcohols, acids and esters in addition to the acetic acid.

At the Nestlé vinegar factory, a small proportion of this malt vinegar is then distilled under vacuum, to produce a colourless vinegar aimed at meeting a specific demand for what is known as "white vinegar". This latter product has a higher acetic acid content than the basic malt vinegar, although it still contains a number of other volatile constituents. Malt vinegar is fed continuously into the distillation vessel with the result that the non-volatile components begin to accumulate and the dry matter content of this residual fraction increases. At intervals the process is stopped, and the **still bottoms** are run off. This co-product, when cooled, is a heavily viscous liquid containing as much as 16-20 per cent acetic acid, and this gives it a strong vinegary smell. However, it is rich in protein and low in fibre, and the NCGD value suggests that the organic matter is wholly digestible. The ash content of approximately 10 per cent is similar to that of other distillation residues such as pot ale syrup, and vinegar still bottoms are a similarly rich source of phosphorus and magnesium. The potassium content is relatively high, but the sodium is at a satisfactorily low level. In order to make this co-product more acceptable as a feed source, it is mixed at the factory with the liquid yeast that was separated at an earlier stage. This blended product is still rich in protein and a highly digestible energy source, but its acidic smell is less pungent and the material is easier to pump.

Malt extract production

Malt extract is the hot water extract of malted barley that forms the basis of an increasing range of unfermented beverages and food products. Its origins lie in the days before history was recorded, but malt extract is known to have been used as a beverage by the early Egyptians (Novartis Consumer Health). However, its nutritional properties were not truly appreciated until the nineteenth century when malt extract was shown to be of particular benefit as nourishment for sick children. Commercial development of Justus von Liebig's fundamental studies was undertaken by George Wander, who pioneered the production of malt extracts (Shenstone, 1895; Wander I). Almost 150 years later, Wander's company, now part of Novartis Consumer Health, remains one of the world's foremost producers of malt extract products. Today, malt extract is widely used as an easily digested ingredient in baby foods, milk-based health drink products and tonics (Wander II), and as a flavouring ingredient for biscuits, confectionary and breakfast cereals. Investigations continue on the value of malt extract's complex mixture of sugars and more complex oligosaccharides as a source of sustained energy, which may be of specific benefit as a component of "sports drinks", particularly those used during endurance events.

Essentially, malt extract is a carbohydrate-rich syrup produced by the low-temperature vacuum-evaporation of wort after mashing and filtration. However, the composition of the grist shows significant differences from those used by the brewer and malt vinegar manufacturer. Whereas, malt vinegar is produced almost exclusively from malted barley, and adjuncts contribute only a modest proportion of the brewers' carbohydrate supply, malt extract producers rely much more heavily on unmalted inputs. Until recently, the standard grist used by the Novartis factory at Kings Langley – the largest malt extract facility in the world, though now under notice of closure – comprised 25 per cent malt and 75 per cent unmalted barley, but those proportions have now been widened even further, to 10:90. Such a low concentration of malt may tax the ability of the product to supply sufficient enzyme activity to solubilise all the starch and protein inputs, although this proportion of malt is used satisfactorily by the grain distilling industry.

Unlike whisky production, where legal restrictions apply, the malt extract industry is able to use additional exogenous enzymes. Whilst augmenting the enzyme supply may be relatively easy, the near replacement of the complex enzyme balance provided by nature for this purpose is a more exacting requirement. Despite the provision of a concoction of amylases, proteases, ß-glucanases and occasionally amyloglucosidases, the complexity of this mixture pales by comparison with the natural content of malt, which is thought to comprise as many as 28 different enzymes (personal communication; P Addison, Novartis Consumer Health). This may partially explain why the spent grains that remain after the separation of the wort sometimes contain more residual starch than would typically be found in brewers' grains. However, in the production of grain neutral spirit, exogenous enzymes may replace the malt entirely and thus this need not be a limiting factor. A further point of difference may be in mashing procedure – although, in malt extract production, the temperature is held at a series of levels to selectively favour the activity of various enzymes, the effective mashing period is believed to be shorter.

At the outset of the process, the high proportion of unmalted barley in the grist has implications for the malt powder that is screened out of the raw material prior to grinding and mashing. Unlike the brewer whose inputs comprise a mixture of malt, syrups and processed cereal fractions such as maize grits, the principal raw material for a malt extract factory is barley. Thus the supply is more closely linked to the farm of origin, and the composition of the screenings reflects this closeness. Barley screenings contain more chaff and straw than would be typically separated from malt powder – a material that has gone through a greater number of cleaning routines.

Apart from the screenings and the "brewers' grains" a proportion of malt extract syrup, or **liquid malt extract (LME)**, occasionally becomes available as animal feed. It is a golden brown, viscous liquid with a characteristic odour and a mild sweet (malted) flavour. Promotional information from Novartis highlights the inclusion in LME of 40 of the 50 trace substances known to be essential for human life, but animal feed specialists would wish to concentrate on the highly digestible carbohydrate content that makes up more than 90 per cent of the feed's dry matter. Composed largely of maltose and oligosaccharides, this material would be suitable for either pigs or ruminants. Pigs may be expected to consume this material readily and it could be included in rations at perhaps 20 per cent of the dry matter, providing adequate fibre is supplied from other dietary sources. However, the use of LME in ruminant diets would need to be restricted to a maximum of 10 per cent in order to avoid the potential problems of a butyric rumen fermentation (Karalazos and Swan, 1976; Swan, 1978).

5 Citrus and Tropical Fruit Processing

Orange and other citrus juices

Citrus are the world's most widely grown fruit and production volumes continue to grow strongly - over the last 40 years, the world production of citrus fruit has increased four-fold. Oranges remain the predominant citrus species, and statistical data recorded by the Food and Agriculture Organisation (F.A.O. 2000) show that orange production continues to grow in step with total citrus volumes. Oranges have made up a consistent 61-65 per cent of the total volume since 1961 – see Table 1.

Table 1: World Citrus Production (mt per annum)

Year	*Total Citrus*	*Oranges*
1961-63	25.5	16.4
1967-69	36.4	23.6
1977-79	54.5	34.4
1987-89	74.0	47.0
1997-99	100.1	63.5

Volumes represent 3-year averages
Source: F.A.O.

F.A.O. records do not reveal the volume of citrus that is currently being processed but Morizzo (Orchard House; personal communication) suggests a figure of approximately 65 per cent. This proportion is in line with the figure given in 1971 by Pollard and Timberlake for the United States crop, and by Hutton in 1987 for the world orange crop. When juiced, up to half the fresh weight is extracted and this leaves the remaining 50-70 per cent available for other purposes. The remaining fraction - comprising the peel, the internal fibrous structures (known as the rag) and the pips – is collectively known as the pulp,. If 65 per cent of the crop is processed, the global figures imply a potential availability of 21-29 million tonnes of orange pulp per annum, together with additional volumes of other citrus pulps.

Although some of this will inevitably be wasted, where small-scale juice production does not justify further processing, in major citrus processing areas the volume of co-product and its efficient handling are of substantial importance. Much of the industrially processed pulp is now dried and exported around the world as an animal feed, but there is a history stretching back more than 70 years of its use as a moist animal feed (Hutton, 1987).

In Britain, citrus fruit are not grown commercially but there is a juice production industry based on imported fruit. British fruit juice consumption has increased by 44 per cent over the last ten years (MAFF, 1998), and orange juice comprises some 77 per cent of the total (Hargitt - British Soft Drinks Association; personal communication). However, the vast bulk of this juice is processed overseas, and only about $2^1/_2$ per cent of British orange juice is squeezed in Britain to supply a premium, fresh juice market. The fresh citrus pulp available to British livestock is derived only from this fraction, but it amounts to an estimated 35-45,000 tonnes per year (A Morizzo, Orchard House; personal communication).

Citrus juices are widely recognised as rich sources of vitamin C in the human diet, although it is notable that the vitamin C content of the pith and peel of oranges, and of grapefruit in particular, is much richer than that of the juice – orange juice contains only 25 per cent and grapefruit juice only 17 per cent of the vitamin C content of the whole fruit (Atkins *et al.*, 1945). This rich supply may be largely wasted on ruminant consumers, which are able to synthesise their own supply of this vitamin. However, there may be occasions when animals under stress have a reduced capacity for the synthesis of vitamin C, and under certain circumstances even ruminant animals have been shown to respond to an additional supply (Gadient and Wegger, 1985). Citrus fruit also contain many other vitamins in amounts that are sufficient to be of dietetic importance; inositol and tocopherol in particular (Rakieten *et al.*, 1947). Citrus fruit are also an important source of bioflavanoids, chemicals whose role is less well understood, though some have been regarded as having a vitamin-like effect (Rusznyak and Szent-Gyorgyi, 1936). Proof of a precise effect may still be lacking, but bioflavanoids have been recognised as being of therapeutic importance in relation to capillary health (Hendrickson and Kesterson, 1965).

The British citrus juice production industry – as distinct from the local dilution of imported juice concentrates – is centred on three factories, in London, the Midlands and in South-East Wales. The citrus growers and packers that supply these factories are expected to comply with Codes of Practice, such as that produced by Orchard House Foods (1997), for the handling of citrus fruit destined for freshly squeezed juice. Essentially the British industry requires fruit that has the lowest possible level of surface bacteria or pesticide residues. Fruit which is not sound and intact, free from damage and/or progressive skin defects, is excluded. However, fruit with superficial blemishes – usually referred to as non-progressive skin defects – are permitted, and this is the sole difference between Class 1 table fruit and Class 2 processing fruit. The need to harvest, select, pack and chill fruit in the shortest possible time is emphasised.

Oranges (including clementines, mandarins, ruby and Florida orange) make up more than 80 per cent of the raw material, with lemons contributing less than 15 per cent, grapefruit approximately 5 per cent, and a small volume of limes also being processed. Typically, a blend of varieties from different countries is used to ensure the production of a consistent fruit juice, in terms of sweetness, acidity, colour and flavour, and the sources are changed progressively as processors follow the orange seasons around the world. From February to July the blend will consist mainly of Caribbean and Mediterranean oranges, whereas from August to January orange juice will be squeezed from Brazilian, South African and other Southern Hemisphere fruit (J Hawkins Sunjuice; personal communication). The predominant variety is Valencia, which accounts for the vast majority of the oranges used for juicing in the UK, and is the preferred variety of juicing companies around the world. Valencia oranges are prized both for their juiciness and flavour, and also because of their lower risk of bitterness. All oranges contain a precursor to a bitter substance known as limonin, and some varieties develop an intensely bitter flavour within 36 hours of squeezing. The relative absence of limonin precursors from the variety selected for juicing may be of significance to the palatability of feed co-products. In contrast to the rapid consumption of fresh fruit juice, the extracted pulp may remain on farms for 1-4 weeks, and even longer when ensiled. However, the greater concentration of limonin in the pips and the peel implies that the co-product is always likely to contain a higher level of this constituent than the juice.

All three factories process a daily batch, which takes in whole fruit in the early morning, squeezes it, packages the product, and off-loads the pulp all in the same day. None of the British supply of citrus pulp is dried, and neither the juice nor the co-product could be fresher when it is delivered to the market. Fresh juice manufacturers can trace their fruit back to the field of origin and, prior to its intake by the factory, the selected fruit will have satisfied a number of safety checks, with particular regard being paid to the use of pesticide chemicals. At the start of every season, all citrus growers must submit their proposals for pesticide use, and the proposed use of any chemicals which are not approved by the EU/UK authorities is rejected as unsuitable. Subsequently, the actual use of pesticides by approved growers, and their strict observance of the required pre-harvest interval, is audited against the proposals. In the country of origin, citrus suppliers make detailed checks of pesticide residues on the fruit at the start of each harvesting season, and continue to make more selective checks throughout the season, in relation to the perceived level of risk. Further testing in the UK builds on the information provided earlier in the chain, and tends to target fruit grown outside the EU - with the previous supply history, and the amount of information available from each supplier, being taken into account.

At the start of each season, citrus growers must submit their pesticide proposals for approval
(courtesy: Orchard House Foods)

In the orchard, correct harvesting procedures have been defined, with emphasis on the avoidance of damaged fruit and soil contamination. The Orchard House Code of Practice also imposes a complete ban on the use of animal manures, which it considers to carry an unnecessary risk of contamination with *E coli* 0157. In the packhouse too, all activities are subject to a Code of Practice, and all fruit is treated in the same manner, whether it is to be processed or used as table fruit. The sequence of events begins with trash elimination and inspection, followed by a disinfectant dip and a fungicidal/detergent wash. The fruit are then rinsed and dried, followed by a further fungicidal/wax application and a second drying. All packhouse treatments which are applied to protect the crop in store, and the waxes which are used to reduce moisture loss, need to be approved by current EU and

UK legislation. After treatment, a second inspection then ensures that only sound fruit are selected for grading, packing and placing into store at the correct temperature.

On receipt at the British processing factory, the fruit is sorted manually to remove any foreign bodies, and to exclude any fruit showing signs of mould development during storage and shipping. These waste materials are collected and disposed of as land-fill – no diseased fruit is added to the citrus pulp at the end of the process. The selected fruit is then subjected to a series of washing and brushing procedures, designed to remove any dirt and mould spores. Washing also removes the waxy coating that was applied in the packhouse. After washing, the fruit is considered to be suitable for the production of juice for human consumption, and the other parts of the fruit are similarly fit for use as animal feed.

The fruit is presented to a large industrial squeezer that sorts the individual pieces according to size. A disc of peel is cut from both the top and bottom of the fruit and the central, pithy core is pushed out. The machine then cuts and presses the fruit, allowing the juice to flow out before discarding the extracted peel in the shape of Chinese lanterns. A proportion of peel lanterns, as they are known in the factory, are finely ground and marketed as "zest" to be used in a range of food factories to add both colour and aroma. The remaining peel, which is of equal quality to that used directly in the food industry as zest, is then recombined with the citrus pith and this mixed co-product of juice production is used in the fresh form as animal feed. At some sites, the pulp is macerated before it leaves the factory, and this is the preferred form for livestock feeding – maceration increases the acceptability of the feed, facilitates its ensilage in an air-tight stack and aids the incorporation of citrus pulp into a complete mixed ration.

In some overseas factories, where the co-product is dried, a slurry of slaked lime is added to the pulp to reduce the hydrophilic nature of the pectins and other fibrous components and facilitate further juice removal. In a study by Pascual and Carmona (1980a), up to 310 litres of juice were removed from a tonne of pulp by the addition of 4 kg of calcium hydroxide, and this was associated with an increase in the dry matter content of the pulp from 21.0 to 24.4 per cent. These authors considered the optimum addition of slaked lime to be 2-3 kg per tonne of pulp, though others have recommended as much as 3-5 kg per tonne. However, slaked lime increases the calcium content of the pulp – 3 kg calcium hydroxide per tonne would increase the calcium content of the dry matter by 0.9 per cent – and this may be of nutritional significance unless adequately balanced. Where the co-product is marketed in the fresh form, treatment with lime does not often occur, and the natural calcium content, of 0.5-0.6 per cent of the dry matter, is typically found.

There is potential for a partial dewatering of the citrus pulp by means of a screw press, and this additional processing would yield both a higher dry matter pulp and a press liquor rich in soluble sugars. Such a procedure was described by Hendrickson and Kesterson (1965), who suggested that the press liquor would typically contain 9-15 per cent dissolved solids, and that more than half of this fraction would be sugars. The volume of released liquor varies with the variety of fruit, its moisture content and the amount of pressure exerted on the pulp. After expression, the press liquor may then be filtered to remove much of the suspended solids and concentrated by evaporation to produce citrus molasses – a co-product syrup with a typical dry matter content of around 71-72 per cent. The removal and concentration of the press liquor is practised in the United States, Australia and possibly other countries, where citrus molasses is marketed as an animal feed to meet a ready demand. Hendrickson and Kesterson (1965) also referred to the recombination of citrus molasses with the pulp to produce a sweetened pulp, akin to the production of molassed sugar beet pulp. Citrus molasses is also used as a feedstock in industrial fermentation processes to yield alcohol, vinegar, yeast, lactic and acetic acids and methane (Askar and Treptow, 1997), and each of these fermentations may be expected to yield a sugar-reduced co-product similar to the condensed molasses solubles produced by the fermentation of sugarbeet and cane molasses. At present, the production of citrus molasses is not a feature of British processing.

Citrus pulp contains approximately 25 per cent pectin (a loose term which covers a wide range of pectic substances – Pilnik and Voragen, 1970) and, with demand from the food industry for the production of jams, jellies and sweets, a proportion of the fruit in Australia, the USA and South America is processed to extract this valuable ingredient. However, the extraction of pectin from citrus pulp requires a complex extraction system, and the quality of the pectin is variable – first grade pectin is claimed to be extractable from lime and lemons but not oranges (Askar and Treptow, 1997). Pectin-extracted pulp inevitably has a different composition from the unextracted co-product, but it does not appear to be marketed separately (Anon, 1992). In common with citrus molasses, the extraction of pectin from citrus fruit is not carried out in British factories, and British citrus pulp continues to include both of these highly digestible fractions.

Gohl (1973) noted that the separation of citrus seeds occurs in some factories, but it is unknown in Britain. Given the great volume of citrus fruit that is processed, the seeds - which comprise some 0-5 per cent of the fresh weight - represent a potentially significant amount. However, Valencia oranges contain only a relatively small proportion of seeds (0.8%), and their separation from this popular variety may be more difficult to justify. A more significant proportion may be obtained from grapefruit processing - the seedy grapefruit cultivar, Duncan, was stated by Hendrickson and Kesterson (1965) to include 4.4% seeds. Citrus seeds represent a potential source of extractable oil that has been used both as a salad oil and in cooking, but the oil content is variable – Hendrickson and Kesterson quote a range of 29 to 45 per cent of the dry seed, while Pascual and Carmona (1980) found oil levels as low as 20 per cent. This wide variation has been linked to differences in species, variety and fruit maturity. Separation of the oil would yield an oilseed meal (or cake) that would have potential for the animal feed market – more particularly for the ruminant feed market. The protein level of the extracted seed was calculated to be 26 per cent by Hendrickson and Kesterson (1965), but these authors also suggested that a protein level of 43 per cent was achievable if the kernel were to be separated from the hull before extraction. Gohl (1973) and Glasscock *et al.* (1950) state that citrus seed meal compares well with many sources of vegetable protein including cottonseed and soyabean meal. However, the relatively high level of acid-detergent lignin in the hull of the seeds may imply a low digestibility, and this would be of nutritional significance. Hendrickson and Kesterson (1965) quote a digestibility value of only 55% for citrus seed meal fed to sheep. Citrus seeds also contain limonin and, unless this component is removed by solvent extraction, it renders the co-product unsuitable as a feed ingredient for poultry and of restricted value for pigs (Driggers *et al.*, 1951).

Citrus pulp also contains a number of volatile components which contribute to its flavour and aroma. These are generally associated with the essential oils which occur in oil sacs located in the outer peel. Peel oils have been expelled in commercial practice in the United States by pressing the peel and, although they comprise only a relatively minor fraction (0.1-0.4 per cent of the dry matter content of the fruit), such oils have a high value. They have found a variety of uses in plastics, pharmaceuticals, soaps, perfumes and household products, in addition to their use as a flavouring ingredient in foods, confections and soft drinks. The separation of citrus oils has a relatively long history in the United States (Kesterson, 1961), but this practice is not a feature of processing in this country. Were it to be carried out, the removal of such a minor fraction would have little effect on the nutritive value of the pulp, although there may be some effect on its attractive aroma. Since the oil is closely associated with limonin precursors in the peel, its removal from some varieties of orange may be expected to enhance the palatability of the remaining fraction.

Fruit salad production

In today's world, food retailers offer a wide range of "ready meals" – food which requires a minimum of preparation before eating. Fresh fruit salad represents one of the dessert options from this range, and it is prepared in a number of food processing factories in the UK – including two which also squeeze citrus fruit for the fresh juice market. With the notable exception of citrus, only Class 1 fruit is used for this purpose, and the mixture comprises a wide array of different species from around the world. The composition of fruit salad varies constantly and it can be tailored to meet specific demands, but at some time it will include strawberries, mangoes, melons, grapes, pineapples, kiwifruit, apples, plums, cherries and a range of orange types such as satsuma, clementine and mandarin.

All fruit can be traced back through the supply chain to the field of origin, and any chemical treatment is closely monitored with checks made before, during and after the crop has been grown. On arrival at the factory, the fruit is inspected and any diseased or damaged pieces are discarded. The selected material is then washed prior to the fruit being processed, and processing inevitably leads to the production of a **co-product fruit salad** containing the following elements ...

- the calyx of strawberries
- the peel of mangoes, melons, pineapple, kiwifruit and orange
- the pips of melon and citrus
- the core but not the peel of apples
- the stalks but not the pips of grapes

Stones removed from mangoes, plums and cherries are not included in the co-product mix, and this fraction is consigned to landfill.

Co-products of citrus processing

Citrus pulp

Citrus pulp is the solid residue that remains after the squeezing of citrus fruit for juice production. The pulp typically amounts to 50-70 per cent of the fresh weight of the original fruit, depending on species, variety and processing technique. The pulp comprises the peel (60-65 per cent), the internal tissues (30-35 per cent) and the pips (0-10 per cent) – these proportions were measured in a study of eight varietal types of citrus pulp by Pascual and Carmona (1980). The predominant citrus fruit is the "common orange" and, although a number of orange varieties contribute to this dominance, the juicing industry is heavily reliant on the Valencia variety. Individual consignments of pulp may be exclusively orange but, occasionally, they will contain a relatively high proportion of either grapefruit or lemon, depending on the volume of orange pulp in the collection system when the production line is switched to one of the minority fruits.

Citrus pulp is an unusual feed in that it has a high fibre content and a high digestibility for ruminant animals. If one adds the content of soluble fibre (pectin) to the NDF fraction, citrus pulp has approximately 50 per cent total fibre in the dry matter, yet the NCGD (neutral cellulase gamanase digestibility) value is around 90 per cent. This high digestibility value obtained in the laboratory is supported by in vivo digestibility values determined in ruminant animals; viz: an organic matter digestibility of 87.2 per cent and a crude fibre digestibility of 93.2 per cent were measured by Fegeros and colleagues (1995). By contrast, both the protein content and its digestibility are low. In a study including eight different varietal types, Pascual and Carmona found a narrow range of protein levels, from 5 (Navel) to 8 per cent (Mandarin), with lemon and grapefruit falling within the range at 7 per cent of the dry matter. Within this band, the protein content of citrus pulp is positively associated with the content of pips. Protein digestibility values have varied within a moderate range of 52.7 to 64.5 per cent (Fegeros *et al.,* 1995; Pascual and Carmona, 1980).

The ash content of fresh citrus pulp is low – the ash values measured in Pascual and Carmona's study varied only from 3.1 to 4.0 per cent. Of particular note is the calcium content, which may be boosted by the addition of slaked lime during processing. Fresh citrus pulp is not usually subjected to such processing and it has a typical calcium content of 0.5 per cent of the dry matter. Dried citrus pulp is commonly treated with lime, however, since this facilitates the removal of water, and its calcium content may be three times as high as that of the untreated material.

There are differences between the pulps of the various citrus fruits, but these are largely restricted to the carbohydrate fraction. Orange pulp typically contains a substantially higher level of sugars than lemon pulp on a dry matter basis, and most types of orange have a lower NDF content. However, the digestibility of the two fruits is not dissimilar – with NCGD values of 91-94 and 88-94 being measured for orange and lemon pulp respectively (James & Son, unpublished data). Grapefruit has a fibre content similar to lemon but there is apparently no information on its digestibility. Differences have also been recorded in the dry matter content of different citrus pulps, with the pulp of lemons and grapefruit tending to be wetter than orange pulp. Although the extent of the difference may vary between factories, and be dependent on processing methods, this factor will have a direct bearing on the financial value of the co-product.

In contrast to the wide difference between the acidity of the juice of lemons and oranges (cf pH values of 2 versus 3-4), there is very little difference between orange and lemon pulps in terms of their acidity; the pH value of both pulps is typically in the region of 3.9-4.0, and the buffering capacity is very similar at 224-228 m.equivs/kg dry matter (James & Son, unpublished data). Notably, the predominant acid of both pulps is not citric but malic (Ting and Attaway, 1971). Both pulps are considerably less acidic than other acidic ruminant feeds such as grass silage, which typically has a buffering capacity in excess of 1000 m.equivs/kg dry matter (McDonald, 1981).

The high digestibility of citrus pulp reflects its potential as a ruminant feed, but a number of farmers have commented unfavourably on its palatability. When offered a choice, they observed that cattle would tend to reject fresh citrus pulp in favour of other feeds, and would selectively reject the pulp in a mixed diet – particularly if the co-product had been offered without maceration, when it is clearly easier for an animal to reject, and for the farmer to notice, large pieces of peel. However, a number of these same farmers reported that acceptability tended to improve when the material had been stored for some time before feeding - most farmers noting an improvement within one to two weeks. A more prolonged improvement was recorded in a dairy cow trial in Scotland (Offer, 1999), where the dry matter intake of a mix of grass silage, citrus pulp and soyabean meal increased progressively from 8.6 kg to 13.0 kg over a six week period. The citrus pulp intake during this period increased from 1.4 to 2.2 kg dry matter. The likelihood that this improvement in acceptability was associated with changes in the citrus pulp, and not adaptation by the cows, was confirmed when the control group of this changeover trial switched to the citrus diet. The new cows, with no previous experience of citrus pulp in the diet, consumed 12.1 kg of the mixed diet in the first week and averaged 11.3 kg over the six weeks spent on this diet.

The factor in citrus pulp responsible for the effect on palatability and intake seems likely to be limonin, an intensely bitter substance found primarily in the seeds and peel, though two other bitter compounds have also been identified. Such chemicals have been classed as "anti-feedant", since in the natural state they deter the browsing animal (Harborne, 1999). The Valencia variety is preferred by juice producers because it exhibits a much reduced development of the bitter principle from a non-bitter precursor, but the limonin content of the pulp is greater than that of the juice. Since on-farm storage of citrus pulp tends to increase the acceptability of this feed to the animal, it seems reasonable to suggest that the bitter compounds may be modified in some way during storage. Although Megias *et al.* (1997) isolated a phenolic-carbohydrate complex from ensiled orange pulp that was capable of inhibiting the digestion of corn silage stover, in vitro, the authors concluded that the total content of this component was too low to pose a problem in most ruminant feeding situations.

The high organic matter digestibility of citrus pulp puts the co-product on a par with rolled barley and wheat as an energy-rich feed for ruminant animals. But there is a marked difference in the source of that energy; cereal grains have a high concentration of starch, whereas the energy value of citrus pulp largely derives from digestible fibre. In this regard, citrus pulp bears a marked similarity to sugar beet pulp and, like beet pulp, its digestion in the rumen is slower than that of cereal grain and more continuous over a long period (Pinzon and Wing, 1975; ADAS, 1989). Compared to starchy feeds, the digestion of citrus pulp is likely to be associated with a much reduced production of lactic acid, and the animal is much less likely to develop a problem of acidosis. Citrus pulp may thus be considered to be a safer feed for animals consuming a highly concentrated, low-roughage diet. High yielding dairy cows may fall into this category, and sheep too, since the group feeding of ewes or fattening lambs often results in an uneven distribution of feed between individuals.

Measurements of rumen fatty acid production (Cabezas *et al.*, 1965; Pinzon and Wing, 1975) confirmed that the replacement of maize grain by citrus pulp tended to increase the acetate/propionate ratio, and this may lead one to prefer the co-product in dairy rather than beef diets. However, citrus pulp can play a part in beef cattle diets, and Jones *et al.* (1942) noted that the co-product can be used in fattening beef diets with no apparent effect on the colour of the body fat, a problem that had previously been of concern. Given the preponderance of acetic acid as the end-point of rumen digestion, it seems likely that the feeding level of citrus pulp should be restricted to no more than 20-30 per cent of the dry matter in the later stages of cattle finishing.

For dairy cows, there has been concern that citrus pulp in the diet may give rise to an undesirable milk taint. Gohl (1973) reported conflicting views, but advised that "grapefruit in particular" should be offered only after milking, in order to avoid milk taint. Hutton (1987) claimed that a milk taint may occur "under certain conditions", but that it had not occurred in his experience of routinely using up to 25 per cent of dried citrus pulp in his dairy supplements. If this can be interpreted as 1.5-2 kg of citrus pulp dry matter per cow per day, it would be in line with the amount of fresh citrus pulp used by Offer (1999) with no reported effect on milk odour or taint. Even higher amounts have been used in a number of trials at the Texas Agricultural Experiment Station (Copeland and Shepardson, 1944), from which it was concluded that no taint was detectable when up to 3.6 kg dried citrus pulp was fed. Probably the longest experience of citrus pulp in dairy cow diets has been recorded by researchers at the Florida Agricultural Experiment Station. When reviewing 60 years of trial work, Harris *et al.* (1982) were led to conclude that citrus pulp (at least in the dried form) has no properties other than its nutrient content which limit its use in dairy cattle concentrate rations. It is notable that these reviewers referred specifically to the use of citrus pulp as part of the concentrate ration because, although citrus pulp may be roughage sparing, they also concluded that it cannot be used in place of the entire roughage allowance.

As a feed for dairy cows, Hutton (1987) reported that dried citrus pulp can replace up to 2kg rolled barley, on an equivalent dry matter basis, without any detrimental effect on milk production or the liveweight of the cow. Concern was expressed that the feed may tend to alter milk composition towards an increase in fat and a reduction in protein content. However, a similar amount of fresh citrus pulp was fed by Offer (1999) without significant effect on either milk yield or quality.

Citrus pulp has been used for sheep in a number of situations. Under the intensive conditions which prevail in the live sheep export trade, sheep were found to adapt quickly to a diet containing 20 per cent dried citrus pulp. The diurnal variation in rumen pH was found to be smaller with the citrus diet than with an alternative diet based on wheat and rice hulls (Hodge *et al.*, 1986; Watson *et al.*, 1986). Fattening trials in Spain have investigated the inclusion of dried citrus pulp at 0-60 per cent of the concentrate feed, and noted satisfactory animal performance up to 30 per cent inclusion, but poorer performance at higher levels of citrus pulp (Pascual and Carmona, 1980b). Fegeros *et al.* (1995) used dried citrus pulp in diets for milking sheep, and found no significant effect on milk yield or composition when citrus pulp replaced 30 per cent of cereal grain.

Although potentially digestible, citrus pulp may be expected to be of lower value in non-ruminant diets, because the energy derived from fibre is absorbed as short-chain fatty acids from the large intestine, and is less well utilised than the digestion products of starch. Hutton (1987) reported a typical usage of 5 per cent in Australian sow diets, which he claimed was in line with European usage at that time. He also emphasised the importance of maintaining citrus pulp in the diet in order to avoid the possibility of palatability problems when the feed is re-introduced. If fibre content were the sole concern, one could point to recent studies with sugar beet pulp that show higher inclusion levels may be possible for both growing pigs and sows (Lee and Crawshaw, 1991). However, there may be other factors that need to be taken into account. Gohl (1973) refers to "substances toxic to swine and poultry in citrus pulp that includes the seeds", and later identifies limonin as a toxic factor for pigs and poultry but not ruminants. Thus, unless pips have been excluded from the pulp, it would seem advisable to restrict the inclusion of citrus pulp in pig diets to a relatively low level.

Fresh citrus pulp has a natural acidity and stores well in the absence of air. The co-product can also be added to grass at ensilage to boost both the quantity and quality of silage. In practice, the proportion of citrus pulp can be varied to meet the needs of individual situations, and it may be sound practice to incorporate sufficient co-product to fill the silo in a single operation and avoid the wastage which occurs when silos are re-opened to add on later cuts of grass. Citrus pulp also enhances the prospects of a good fermentation. The pulp contains a number of organic acids (Ting and Attaway, 1970) – particularly malic and oxalic, but little citric – and these have an immediate impact on herbage pH, which reduces the amount of fermentation acid needed to preserve the crop. The reduction in pH stimulates the activity of naturally occurring lactic acid bacteria so that fermentation occurs more swiftly, leaving less time for the sugars to be wasted by respiration in the silo. And for good measure the citrus pulp adds extra sugar, which ensures that even the most difficult crops can achieve a satisfactorily low pH. Dried citrus pulp offers a further advantage, when stored in conjunction with other feeds, in that its absorptive nature restricts the loss of effluent and associated nutrients. This effect is of particular value to silage makers operating in environmentally sensitive areas, but the co-product can also be used with other moist feeds – such as brewers' grains – to reduce the effluent risk and nutrient loss from such materials.

Citrus molasses

Citrus molasses is a syrup produced by the concentration of juice released from citrus peel, usually as a result of treatment with lime and subsequent pressing. Since oranges represent the predominant fruit processed by the juicing industry, citrus molasses is largely derived from orange processing. The press juice, as this material is known prior to concentration, is rich in sugars – the sugar content is claimed to be 15 per cent higher than that of the fruit juice (Hendrickson and Kesterson, 1965). After separation from the peel, the press liquor is screened to remove the larger particles of pulp and sterilised by heating. As well as the destruction of undesirable micro-organisms, heating also drives off the peel oil and encourages the precipitation of calcium salts and the flocculation of other suspended matter. The clarified liquor is then concentrated by evaporation, and this may be performed in more than one operation with further screening to remove more suspended solids that may otherwise increase the viscosity of the molasses.

The typical dry matter content of citrus molasses, at 71-72 per cent is marginally lower than cane or beet molasses, but the sugar contents are roughly similar at 60-65 per cent of the dry matter. The protein content of citrus molasses is relatively low, and much closer to the protein level of cane than beet molasses. However, the ash content of citrus molasses is significantly lower than either of the other two molasses types, and this is particularly reflected in the potassium content; (cf 1.5, 3.9 and 4.9 per cent of the dry matter) for citrus, cane and beet molasses respectively. Viscosity can be a problem with all types of molasses, particularly in cold weather, and citrus molasses can give additional difficulty in store unless stirred regularly.

Citrus molasses has found ready acceptance by beef and dairy cattle, and was shown to be more palatable than cane molasses in comparisons by Chapman *et al.* (1953) and Kirk *et al.* (1956). Its feeding value was confirmed by these authors and by Baker (1950), who replaced half the ground, snapped maize in a steer fattening ration. Pigs needed 3-7 days to accustom themselves to the flavour of citrus molasses in work by Cunha *et al.* (1950), but thereafter the co-product satisfactorily replaced up to 40 per cent of maize in the diet.

Fruit salad

Co-product fruit salad is the mixed material that becomes available when a wide range of fruits, imported from a number of countries around the world, are processed in the production of ready-to-eat fruit salads for the retail market. The co-product comprises the peel, pips and stalks from a number of mainly tropical and sub-tropical fruit together with the cores of apples. The non-selected fraction of stone fruit such as peaches, plums and cherries is excluded.

Fruit salad is an attractive, succulent feed with a relatively high moisture content. Its composition is variable depending on the ingredients used and their proportion. Since it mainly comprises the peel and the pips of the fruit, co-product fruit salad may be expected to have a higher content of protein, fibre, ash and oil but a lower content of sugars. Although the co-product is acidic, it is less so than the internal flesh of the fruit from which it is derived.

Fruit salad is essentially an energy feed, which can be used in both ruminant and pig diets. However, its variable content and physical form may encourage some selectivity and ideally, the co-product is best fed as part of a mixed diet, where it would provide a succulent alternative to cereal grain. In commercial use, 10-14 kg fruit salad per head per day has been fed as part of mixed diet for dairy cows. It has stored well on the farm for periods up to a month, and has proved to be a palatable ingredient that the cows were seen to select from the diet (M. Bridewell; personal communication).

6 Distilling

The origins of Scotch whisky may be lost in the mists of time, but the technique of spirit distillation is believed to have begun in Asia as far back as 800 B.C. Closer to home, the distillation of an alcoholic beverage was certainly practised by the Ancient Celts, who considered their product to be a gift from God, with the capacity to revive both tired bodies and minds. Thus the Scots do not claim to have invented the distilling process, merely to have "perfected it" (SWA). The earliest documented record of whisky production confirms that the product was distilled more than five hundred years ago, and enjoyed at the highest level of society. Aqua vitae, as it was known at the time, is believed to have been served to King James IV on a visit to Inverness in 1506 (SWA, 1997). By 1644, whisky had become so popular that the Scottish Parliament levied an excise duty on the spirit. However, the collection of tax on the Scotsman's native drink was never easy, and it proved even more difficult when the English revenue staff assumed the responsibility following the Act of Union in 1707. Throughout the next century much of the whisky that was consumed had been produced in the innumerable illegal stills that appeared to spring up as quickly as others were closed down. The scale of the activity may be gauged from excise records that confirm the confiscation of as many as 14,000 illicit stills per year by the early 1820's. Some order was finally restored to the embryonic industry in 1823, when the Excise Act was passed and at last it became profitable to produce whisky in a legal manner (SWA).

Vodka and gin producers may dispute the Scotsman's claim to the perfect spirit, though it is notable that much of the alcohol distilled for the British production of these drinks is produced in Scotland. Vodka has the longer history; the first record of its production being in ninth century Russia, though there are rival claims of its earlier production in Poland. By the 14th century, vodka had been recognised as the Russian's national drink and, within the next 200 years, it had become the national drink of Poland and Finland too. The spirit was first exported from Eastern Europe around 1500, but a wide range of flavouring materials were incorporated and the production of a standardised quality vodka was not achieved for another 400 years. Its production in the Western World can be traced back to 1934 and the meeting of two Russian emigrés in Paris. The Smirnoff family business in Moscow had been confiscated after the Russian Revolution and it was now revived with the opening of the first vodka distillery in France (GVA, 2001).

Gin has a much shorter history than vodka and the earliest confirmation of its production comes from 16th century Holland. British soldiers fighting "bleak campaigns in the Low Countries" during the Thirty Years War became familiar with its restorative effects, and supplies of gin reached Britain from Holland in the 1600's (Bayley, 1994). The distillation of British gin was encouraged during the reign of Charles I and later by William of Orange, who came to the English throne from Holland in 1689 and promptly raised the excise duty on French wine and brandy. By the 1730's, gin was available in over 7000 shops in London alone, and gin consumption exceeded that of beer and ale. Drunkenness became a scourge of the common people as the distillers' output rose twenty-fold over a 45-year period, amounting by the 1730's to the annual equivalent of 63 litres per adult male in the London area. At that point the Gin Act was brought into law; this raised the excise duty five-fold and banned the sale of gin from unlicensed premises, and in quantities of less than 2 gallons (9 litres). The Act provoked rioting on the streets, and open disregard for the licensing law, without achieving its intended purpose – within six years, gin production (and consumption) had increased by a further 50 per cent. The Gin Act was repealed in 1742, and a new policy was introduced that gained more widespread acceptance amongst both producers and consumers (GVA, 2001). It would be a further 80 years before a similar peace broke out North of the Scottish border.

Malt whisky

Once the legal hurdle had been overcome, a large number of malt distilleries sprang up in the North East of Scotland and this area remains the heart of the industry today. Distilleries were also established in other parts of the Highlands and in the Western Isles, where over the last one hundred and fifty years they have contributed much to the economy of those areas. The industrial revolution of the mid-nineteenth century led to the development of domestic stills into factory-scale operations, and those distilleries employed a substantial proportion of the local labour force. But malt whisky continued to remain a specialist item, produced by traditional practices from traditional ingredients, and created in different guises at a hundred different locations. Much of the distinctive character of malt whiskies is provided by the local water supply, and in some cases by the peat that is burnt during the kilning of the malt. The importance of such local factors means that it would be unthinkable to consider alternative locations.

When attention is focused on the co-products of malt whisky production, those local factors must be seen as a disadvantage. The distance of malt distilleries from large concentrations of animal production constitutes a significant logistical problem to the continued clearance of the distillery and the supply of animal feed to farm. The co-products comprise a solid fraction, the draff, consisting of the extracted grain residues and a nutritious liquid, known as pot ale, that remains after the distillation of the spirit. From each individual distillery the co-product output may require only a modest marketing operation but, collectively, the total yield from the industry is substantial, and far outweighs the feed requirements of local livestock farmers. The location of whisky production sites is shown in the map at the front of this book. Thirtyseven malt distilleries span the Highlands from Orkney in the North to Campbeltown in the South, with an additional concentration on Speyside, where 49 distilleries are crowded into a narrow corridor stretching North-East of Aviemore. The 7 Islay distilleries are considered to be a group in themselves, though the Western Isles also include distilleries on Jura, Skye, Arran and Mull. In marked contrast to the concentration in some parts of the Highlands and Islands, the Scottish Lowlands boast only four malt distilleries and these are widely spread from Dumbarton to Wigtown to Glenkinchie, which lies to the South-East of Edinburgh. Only this latter group could be considered to be located within easy reach of a significant livestock feed market. The distant location of most of the other 93 malt distilleries is a dominant factor when deciding the optimal utilisation of the co-products that are an inevitable accompaniment to whisky production.

Lagavulin is one of the seven distilleries on Islay
(courtesy: Guinness UDV)

The annual output of malt whisky has varied widely over short periods of time and, with it, the volume of associated co-products. The Scotch Whisky Association records for the last 30 years show a two-fold variation between the peak year of 1974 and the much leaner years of 1983 and 1984 (SWA, 1999). Over the last decade production has been more consistent, though by no means stable, with the yield in 1997 exceeding those in 1995 and 1999 by more than 20 per cent. Table 1 shows a summary of malt whisky production and the potential output of associated animal feeds since 1969 - all the figures are presented as 3-year means. The alcohol figures have been extracted from statistics published by the SWA and may be expected to be wholly accurate. The potential yields of moist and liquid feeds have been calculated from a model proposed by R T Pass of United Distillers and Vintners (personal communication) – see footnote to Table 1. An earlier model, covering the entire cycle of activities, including water inputs and gaseous outputs as well as whisky and animal feed production, was published by the Malt Distillers Association (MDA, 1994). (N.B. PAS is a concentrated form of pot ale and is discussed later.)

Table 1: The annual production of malt whisky* and potential yield of feed co-products 1969-1999

Year	*Malt whisky** m LPA*	*Moist draff*** '000 tonnes*	*PAS '000 tonnes*
1969-71	151	376	104
1974-76	187	468	129
1979-81	164	410	113
1984-86	102	257	70
1989-91	182	455	125
1994-96	158	396	109
1997-99	179	448	123

Source: SWA (1999)
*Malt whisky production is expressed in million litres of pure alcohol
**Potential volumes of the solid and liquid animal feed co-products have been calculated from whisky statistics by use of the formula: 1 tonne malt yields 400 litres alcohol + 1.0 tonne draff + 0.27 tonnes pot ale syrup (PAS).
N.B. Much of the co-product volume shown in the Table is dried before being marketed to the animal feed industry.

It should be noted that the PAS volumes given in Table 1 represent the potential of the malt distilling industry. Since a proportion of the pot ale is not recovered as animal feed, but disposed of in other ways, the actual amount used as animal feed will be less. A detailed assessment of the pot ale used in feed has been made by

R T Pass in the chapter on distillery feeds in a forthcoming book (Pass and Lambart, 2001).

The calculated volume of moist grains from malt whisky production represents a vast quantity of feed. The mean annual volume for the 1997-99 period would be capable of providing a typical allowance of 9 kg per head per day to 249,000 dairy cows throughout a 200-day winter. At an average herd size of 88 cows (SAC, 1997), that implies a requirement for 2828 dairy herds to consume the total amount of malt grains. This number greatly exceeds the total number of dairy herds in Scotland, which are currently estimated to be in the region of 1800. Even if those numbers were in balance, there would still remain the formidable problem of a 200 mile discrepancy between production sites centred on Speyside and the potential dairy market of South-West Scotland. Malt grains can of course be fed to beef herds and sheep, and there are greater concentrations of these animals than dairy cows in North-East Scotland. But these classes of stock are less productive, and their nutrient requirements are consequently much smaller than those of the dairy cow. Moreover, the needs of beef cows and sheep are largely met by forages grown on the home-farm and thus, although moist grains can be made available at a competitive price, the proximity of local beef herds and sheep flocks does not remove the imbalance between supply and demand.

Processing

The ingredients permitted for malt whisky production are strictly defined by law; barley, malt, yeast and water - nothing else is permitted. The prime ingredient is malt, and the industry traditionally germinated its own barley and used much of the malt it produced in the green (unkilned) form. In modern times, the distillers largely purchase their malt in kilned form from specialist maltsters. This frees them to concentrate on their core business, and the industry gains the benefit of rigorous quality checks imposed by the maltster on his grain suppliers to safeguard the raw material against moulds, insects and a wide range of potential contaminants. Unlike the brewer, the distiller has no interest in the colour of the malt and he can define his requirements more narrowly. Malt distillers essentially require a low-nitrogen malt, capable of producing a high yield of fermentable carbohydrate.

At the distillery, the selected malt is crushed into a coarse meal, known as the grist, and transferred to a large, circular mash tun where it is infused with hot water and held at a temperature of around 63°C in a process known as mashing. The aim is to provide the optimum conditions for the enzymic conversion of starch into the simple sugars that can be assimilated by yeast during the fermentation stage. Mashing is a relatively rapid process and, when complete, virtually all the available carbohydrate and some of the protein will have been solubilised. The liquid phase, containing some 15 per cent of sugars, is then filtered through the bed of spent grains, and this solution, known as the wort, will constitute the feedstock for the alcoholic fermentation. The extracted malt residues are sparged to remove the remaining sugar and then diverted from the whisky production line to become available as animal feed. These distillers' grains in their fresh form are popularly known as **draff**.

Within the distillery the wort is cooled to a temperature of 25-30°C, and transferred to a large fermentation vessel, known as the washback. The distillers' wort, unlike that of the brewer, is not boiled, and thus some enzyme activity continues in the next stage – albeit at a slower pace at the lower temperature. Carefully selected strains of yeast (*Saccharomyces cerevisiae*) of good viability are added, all of which are obtained from outside sources. In a further variation from brewery practice, no yeast is cropped from the washback to be re-used in the next fermentation batch. Yeast may be added in a number of different forms but it will take some 12-20 hours before it begins to assimilate sugars. However, over the next two days or so, these micro-organisms will multiply and produce a vigorous fermentation in which all of the available sugars will be metabolised, with the eventual production of alcohol and carbon dioxide. Although the conversion of sugar to alcohol is a well-recognised process, it is effected through a complex series of steps involving sugar phosphates, sugar acid phosphates, pyruvic acid and acetaldehyde (West and Todd, 1957). This complexity emphasises the fact that fermentation is a live system, and that consideration must be given to the nutrient requirements of the yeasts if they are to perform to best effect. It is also important to appreciate that the conversion process is specific, both to the yeast micro-organism and to a limited range of simple sugars. Contamination by bacteria or other fungi would result in the fermentation of sugars by a number of different pathways and lead to a variety of different fermentation products.

At the end of this stage, the fermented liquid is referred to as the wash and it comprises a dilute (8%) solution of alcohol, together with other fermentation products, yeast debris and some unfermentable material. The wash will be distilled twice in large copper pot stills to separate the alcohol from other components, and the distilled spirit will subsequently be transferred to oak casks for maturation. The spirit will remain in the casks for a minimum of three years before it may be marketed as a "single malt" or blended to produce one of a range of branded whiskies.

After the primary distillation, the liquor remaining in the still is the pot ale, and it consists of all the non-volatile components of the wash. The yeasts that proliferated during the fermentation stage make up a substantial proportion of the solid component, and they give it a significant protein content. Up until the early years of the 20th century this material was used as animal feed but, with a dry matter content of no more than five per cent, its distribution in this form was uneconomic. As production volumes grew, pot ale was turned into fertilizer or carried by pipeline for disposal at sea. However, such disposal routes were not available to all, and both economic and environmental considerations dictated that an alternative processing method must be found. This led to the installation of evaporators in distilleries and to the concentration of this residual liquor into a more valuable **pot ale syrup** (PAS). At different locations, varying degrees of evaporation were achieved before the increasing viscosity of the syrup limited further water removal. Beyond this point, the syrup became too stiff to pump, and consequently too difficult to be marketed and used as a liquid – particularly during the winter months. The acceptable limit for the viscosity of this syrup usually occurs at a dry matter content of 40-50 per cent, but factors other than dry matter content also play a role. A high proportion of suspended solids in the PAS impairs the flowability of the syrup, and reduces the dry matter content at which the material can be marketed as a liquid. Soluble components of the original cell walls, and particularly the ß-glucans, may also increase viscosity, but their effect in PAS is far less than that experienced by the grain whisky industry when evaporating its equivalent distillation residue.

The residual liquor following the second distillation is known as spent lees. This liquid is very largely water, though it does contain some volatile components of the wash other than alcohol. The dry matter content of spent lees is very low and its nutritive value is negligible. It has traditionally been disposed of on land or to sea or, where effluent treatment is available, it is subjected to a biodegradation process prior to the discharge of a low B.O.D. liquid into the drains.

N.B. The copper stills are a traditional and characteristic feature of malt whisky production, and are used to remove sulphurous compounds from the liquor by combination with dissolved copper. However, this leads to a substantial increase in the copper content of the pot ale that remains after distillation. Whereas barley, the principal raw material of malt whisky, has a typical copper level of 4 mg/kg DM, pot ale has a copper level in the region of 100 mg/kg DM (MAFF, 1990; Black *et al.,* 1991). Although such a high copper content poses no problem to adult cattle, some breeds of sheep are particularly susceptible to copper toxicosis (ARC, 1980). Thus PAS, and any blended co-product containing PAS, have generally been regarded as unsuitable for feeding to sheep. However, the bio-availability of the copper in PAS is very low, which implies that sheep will not accumulate potentially harmful amounts of PAS copper as quickly as they would with a similar intake of the element from other sources (Suttle *et al.*, 1996). This finding has opened up the possibility that PAS and its derivatives may be used for sheep under specified conditions.

Copper stills are a traditional feature, and a range of distillers' feeds have a high copper content
(courtesy: Guinness UDV)

Drying the co-products

By the early 1900's, although draff continued to find a ready market on neighbouring livestock farms, the volume of pot ale had overwhelmed what local demand there was for such a liquid. The problem was felt most acutely on Speyside, where the industry was most concentrated, and in 1904 the distillers formed a joint company to tackle this problem for the common good. The Combination of Rothes Distillers Ltd (CRD) built a drying plant that converted much of the local pot ale supply into a dry powder, and it was sold as a fertilizer under the name Maltosa. That arrangement eased the situation considerably, and for the next fifty years the industry was able to concentrate its attention on whisky production. But as production grew during the 1960's, there was insufficient demand for the increasing volumes of the solid feed co-product, and CRD took the decision to build its first dark grains plant. This was a clever idea because it not only solved the immediate problem of the unwanted draff, but it provided the opportunity to improve the value of the pot ale too. Since that initial installation, other drying plants have been built on Speyside and, at present, there are three large drying facilities that handle all the co-product that cannot be placed at an economic cost on the local feed market.

When the co-products arrive at the dark grains plant, the draff is firstly screw-pressed to remove as much of the juice as possible prior to drying. An efficient press will remove more than half the water that is present in the spent grains supplied by the distilleries and lift the dry matter content from 22 to 35-40 per cent dry matter (Agnes Peters, CRD; personal communication). At CRD, the pressed draff liquor has a total solids content of about 4 per cent, but more than half of this is present as a fine suspension. Viscosity problems are inevitable if such a material were dewatered in this form and the liquor is thus filtered to remove a proportion of this suspension. The particulate matter is returned to the wet draff for further pressing, and the filtrate is added (1 part to 10 parts) to the pot ale. This liquid mixture is then evaporated in an energy-efficient process that results in a syrup with a dry matter content of 45-50 per cent, and forced-circulation finishers are employed to ensure the removal of the maximum amount of water. The pressed draff is passed to a paddle mixer where it is blended with the syrup, and the moist blend is then rapidly dried in a stream of hot air. The resulting dried material, known as **barley (or malt) distillers dark grains**, is extruded through a pellet mill and then cooled. The pellets are screened and fine particles are returned to the feed hopper for further pelleting. Distillers dark grains, after screening, are marketed to the animal feed market and they find a ready demand in the livestock production centres of England and Scotland – well beyond the local market that could be reached at an economic cost by the undried feed materials.

In other parts of the Highlands, and in the Western Isles, the discharge of pot ale to sea is still permitted from coastally-sited distilleries, and this option is more economical than the installation of dark grains drying plants. Environmental pressures may eventually force a change in this policy. At other sites, the evaporation of the liquor has been the only viable option though, at times, low market demand and lack of profitability have led to a build-up of surplus stocks of PAS. However, the commissioning of additional drying capacity on Speyside during the 1990's relieved the pressure on the liquid feed market, and brought supply and demand into better balance. The disposal of draff still presents a problem in some areas and particularly on Islay, where distilling activity is highly concentrated but livestock farming is pursued on a much more limited scale. However, drying the co-products is not feasible because of the problems of energy supply on the island. Thus here, as in other remote areas, draff continues to be fed in its fresh form, and it is commonplace for the feed to constitute the major portion of the animals' diet.

Grain whisky

Grain whisky production began more than 300 years later than malt distilling, triggered by the development of a continuous still, and from the earliest days this was an industrial-scale operation. An improved version – the Coffey still - came on the scene in 1831, and use is still made of it to the present day. The industry's much greater scale can be judged from the fact that the total production from 8 grain distilleries exceeds the output from 97 malt distilleries by 39 per cent (mean value for a ten year comparison, 1990/99). Grain whisky's location in the central lowlands of Scotland can be explained by a number of reasons, though good communications and a ready outlet for the distilleries' co-products and wastes contributed to the decision. Thus from the outset, the new industry would be less afflicted by the geographical difficulties of supplying a relatively low-priced co-product to a distant market.

The major difference between grain and malt whisky production is the use of additional cereals – whereas malt whisky production uses malted barley as the sole source of fermentable carbohydrate, maize or wheat comprises 85-90 per cent of the cereals used in the production of grain whisky. In this respect, the raw materials of a grain distillery bear some similarity with those used in brewing, where the malt is "extended" by the addition of cereal adjuncts to supply extra carbohydrate, though grain distillers "extend" their malt to a much greater degree. A closer comparison could be drawn between the grain whisky and malt extract industries, where the proportion of malt in both cases may fall as low as 10 per cent. However, a crucial difference remains that distinguishes the two processes in the mash tun - whereas the carbohydrate-rich malt extract is released by a range of exogenous enzymes, whisky production must rely entirely on enzymes supplied by the malt. This legal requirement puts the grain distillers' malt under greater pressure; in a grist that may contain only 10 per cent malt, the enzymes have to do ten times as much work as in a malt distillery. Thus the grain distillers' malt specification is quite different – both a high-nitrogen content and high enzyme potential are stipulated, and the malt's value as a source of fermentable carbohydrate, if not incidental, is certainly a secondary issue.

At Cameronbridge, on the Firth of Forth, grain whisky began as a secondary activity in a flour mill, though the new recruit was quickly to overhaul its senior partner. The flour mill origin dictated that wheat would initially be the principal raw material, though the distillery has relied on maize when this has been more economic. By way of contrast, the North British distillery in Edinburgh began as a maize-based operation, served by ships carrying grain from North America and Europe into the port of Leith. Over the last hundred years, the distillery's reliance on maize has varied in line with its price relative to that of wheat, though the calculation also takes into account the higher extraction rate that can be achieved with maize. In the recent past, the grain distillers have also had to cope with the vagaries of Intervention Board refunds, and to make their choice of cereal on the basis of a figure that varied on a daily basis in line with the difference between EU and world grain prices. Following the convergence of British cereal prices with the world price, this complication has now been removed.

Like its malt equivalent, grain whisky must be matured in oak casks for a minimum of three years, but, following that period, it is typically blended and marketed at a much younger age than the malt product. Despite this major difference in marketing, the pattern of production of both whiskies (into warehouse) has been very similar (SWA, 1999). Thus 1973 and 1974 were peak production years, and the low point came in 1983. Since that time the production of grain whisky has gradually increased, though the 1973 peak has been surpassed only in 1997. Table 2 provides a summary of production statistics since 1969 – in common with Table 1, the figures are presented as 3-year means. The calculation of co-product volumes is less easy than in the malt industry, where the raw material and the processing are basically similar in all the distilleries. This is not the case in the grain whisky industry, where major differences occur between plants, and a greater variety of co-products is produced. The whisky production data have been taken from SWA statistics (SWA, 1999), while the co-product data in Table 2 have been calculated from detailed estimates made for 1997 by R W Pass (United Distillers & Vintners; personal communication) and extrapolated to other years in proportion to alcohol production. The co-product volumes include moist and dried materials, both of which have been converted to a dry matter basis – assuming dry matter contents of 90% for dried and 29.5% for a weighted average of the moist feeds.

Estimates of the production of co-products for the year 2001 have also been made by Pass (personal communication) at 99,000 tonnes of dried and 119,000 tonnes of moist material, equivalent to a combined total of 124,000 tonnes of dry matter. This volume is substantially lower than the amounts for the last decade, which are shown in Table 2, and thus the variability in feed production by the grain distilling industry continues as it has done in the past.

Table 2: The annual production of grain whisky* and co-product feed grains 1969-1999

Year	*Grain whisky** *m LPA*	*Co-product '000 tonnes DM*
1969-71	217	139
1974-76	224	144
1979-81	217	139
1984-86	157	101
1989-91	228	146
1994-96	235	151
1997-99	257	165

*Grain whisky production is expressed in million litres of pure alcohol
The figures include evaporated spent wash that is recombined with the solid co-product and dried, but not the smaller volume that is sold separately in liquid form.

Despite variability, the volume of dried grains from the grain distilleries of Southern Scotland can be easily absorbed into the 4 million tonne British ruminant compound feed market (MAFF, 2001), or into the equally large volume of dry feed materials sold direct to farms. The moist feed volume, which must be marketed more locally, is more significant – and so is its variability. The 1997 volume of moist co-products equates to more than 1 tonne per annum for every dairy cow in Scotland; i.e. 5 kg/head per day for the duration of a 200-day winter. On its own, this would be a challenging target for any feed merchant, but it should be noted that this feed supply is additional to the moist feeds supplied by the malt whisky industry. Thus the importance of drying facilities that extend the market for the co-products of both sectors of the whisky industry cannot be overstated.

Prior to their inclusion in the mash tun, unmalted cereals may need to be cooked in order to hydrate the starch and facilitate enzymic breakdown. Maize invariably needs to be cooked, though it is first milled to a coarse grist in order to expose the starch before it is inserted into the pressure-cooker. The maize may be held at an effective cooking temperature of 150°C for half an hour, or at a lower temperature for a more prolonged period. During this time the starch granules are gelatinised; they swell and break open, and render the starch more accessible to the malt enzymes that will transform it during mashing into simple sugars. Wheat flour may be used without pre-cooking because the temperature at which the grist will be mashed is sufficient to bring about a gelatinisation of wheat starch. However, the breakdown of uncooked wheat starch may be very slow, and distilleries often resort to cooking in order to speed up the solubilisation of the starch and its subsequent conversion to fermentable sugars. Where whole (unground) wheat grain is used, the starch is insufficiently exposed and, in this situation, cooking is essential to bring about the gelatinisation of the grain and facilitate enzymic attack.

At two of the eight grain distilleries (Dumbarton and North British), the wort is filtered through the spent grains after mashing, in much the same manner as that employed by the malt distillers, while no filtration takes place at this point in the other six distilleries. Although the solid co-product that may be removed at this stage is also referred to as draff, the grain distillers' spent grains are of a different quality to those produced in a malt distillery, because they largely consist of wheat and maize residues rather than those of malted barley. However, none of this material becomes available in moist form for animal feeding; it may be pressed to remove as much water as possible or simply dried in a large steam-heated rotary drier. Where it is available, the press liquor is recycled into the mash tun to be used in the extraction of more grain. The separated wort is cooled and yeast is added as it is pumped into large wash-backs, where fermentation of this sugar-rich solution will take place over the next 48-72 hours. The resulting wash is continuously fed into the still, and the end-products are continuously removed: the alcohol is distilled off and associated co-products collected. The alcohol is then casked for maturation and most of it is blended to produce a range of branded whiskies, though a small quantity is marketed as single grain whisky.

The residual liquor remaining after distillation of the spirit is known as spent wash and, like the malt distiller's pot ale, it is a potentially nutritious liquid in highly dilute form. However, this co-product is different in character in that it is composed of dissolved and particulate matter from wheat and maize as well as the malt and yeast residues. Where wheat is the principal grain input, this difference is often associated with viscosity problems when the spent wash is evaporated. The problem is largely caused by the presence of xylans, arabinoxylans and ß-glucans, which are types of non-starch polysaccharide (NSP) found in the endosperm cell wall of wheat and barley, but to a much smaller extent in maize. During malting, enzymes are produced that break down the long-chain ß-glucans of the barley, converting them into smaller molecules that can yield additional amounts of fermentable sugar. But these natural enzymes are inactivated at the temperature of the mash tun, and thus the grain distiller's substantial reliance on unmalted cereals means that the xylans and arabinoxylans that are present in wheat are not broken down. In fact, these long-chain molecules tend to aggregate and the viscosity of the solution increases (Rhône-Poulenc, 1997A). This aggregation increases at the higher temperatures that prevail in the evaporator and the viscosity problem is exacerbated by the loss of water from the evaporating wash. The severity of this effect varies between seasons, in line with variation in the NSP content of the grain, but it is a major factor that limits the effectiveness of the evaporator. Technology can be used to ease the problem and commercial sources of heat-stable ß-glucanase enzymes - which have additional xylanase activity - are available that will break up the long chain molecules and reduce viscosity. Additional benefits are also claimed for ß-glucanase enzymes, in terms of reduced fouling of the evaporator, better heat transfer and faster evaporation (Rhône-Poulenc, 1997B), and these very practical considerations help to explain why such enzymes have been widely used.

At the North British Distillery, before any effort is made to concentrate the spent wash, it is spun in large decanter centrifuges to reduce the high proportion of suspended solids. A solid fraction, known as dreg, is separated from the dissolved solids of the liquid centrate and then dried to 80-85 per cent DM content. The centrate is evaporated in a series of steps that increase the dry matter content from 2.5 to 45 per cent and yield a smooth, brown syrup. The two spent wash components are then recombined and mixed with dried (grain distillers) draff, before being subjected to further drying. At the completion of these various procedures, there is eventually only one co-product, known to the distillers as dark grains, though the cereal origin may also be identified. The dried co-product may be pelleted or sold as a coarse meal; both types are used by the animal feed industry for feeding to cattle and pigs throughout the UK. To distinguish between the various types of dried distillery co-product, the grain distillers' materials are usually referred to as Distillers Maize and Distillers Wheat, although these popular terms are less than definitive. The co-product of mixed origin is often more fully described as Grain Distillers Dark Grains. and other proprietary names are also used.

At the other six grain distilleries, the draff is not separated from the wort after mashing, and the entire contents of the mash tun are transferred after cooling into the washbacks. The mass of extracted grain solids and liquid extract are inoculated with yeasts at approximately 1 per cent of the volume, and fermentation proceeds over the period of the next two days. On completion of the process, the whole wash is transferred to a continuous still where the alcohol is vapourised and removed leaving a residual spent wash. Only at this late stage, are the solid grain and malt residues, together with the debris of inactivated yeast, separated from the liquid phase.

At five of these distilleries, decanter centrifuges are used to effect a separation of the two streams and, at three of them (Invergordon, Port Dundas and Strathclyde), the liquor is evaporated before being re-combined with the solids, and the whole co-product output is then dried. Like that from Dumbarton and North British, the dried co-product is marketed as grain distillers dark grains – mostly of wheat derivation. At Cameronbridge and Loch Lomond, the two fractions remain separate and are not dried; their principal co-product being a solid that is marketed in moist form as yeast-enriched grains. At 25-27 per cent dry matter, these grains are drier and of notably higher nutritional value than malt distillers' draff, and to distinguish this superior co-product from draff, the soubriquet "super" is incorporated in the name. Thus **Supergrains** are available from Cameronbridge and **Loch Lomond Supers** from the Loch Lomond distillery. At Cameronbridge, the decanter settings are aimed at maximum solids removal and the release of a clear juice that is approved for disposal at sea. At Loch Lomond, the liquid will be concentrated and marketed separately as an evaporated spent wash, known as **Loch Lomond Gold**, for both pig and ruminant feeding.

This gives Loch Lomond greater flexibility in its decanter settings, allowing a proportion of the solids to be switched between the liquid or the solid fraction. Typically, this flexibility is used to produce a solid co-product at a slightly higher dry matter content than is achieved at Cameronbridge; viz: 27 v 25 per cent. However, the corollary of this adjustment may be a tendency for greater viscosity in the syrup fraction produced at Loch Lomond. Viscosity has already been noted as a potential problem in evaporated spent wash (ESW), which is also available in limited quantity from some of the grain distilleries where most of the co-product is dried. For this reason, ESW has a typical dry matter content of only 34-38 per cent, significantly lower than that of pot ale syrup.

At Girvan, the mashed grains are not filtered and a spent wash is produced comprising both liquid and solid components. But, in place of the decanter centrifuges, filter presses - incorporating a series of very fine polypropylene cloths – are used to recover the suspended solids from the residual liquor. The resulting moist solid contains virtually all of the insoluble material in the spent wash and, at approximately 35 per cent DM, it has a relatively high dry matter content. This moist solid is supplied, under the name **Vitagold**, to the animal feed markets of Scotland, Northern Ireland and the northern half of England. The liquid separated from these processes at Girvan is discharged to sea under permit.

Grain neutral spirit

Grain neutral spirit (GNS) is the alcohol base used by most gin and vodka distilleries. Although these spirits are produced in smaller quantities than malt and grain whisky, their combined volume continues to grow. Table 3 presents data collected over the last 18 years by the Gin and Vodka Association of Great Britain (GVA).

Table 3: The annual production of gin and vodka in Britain 1978-2000

Year	*Gin*	*Vodka*	*Total*
1978-80	42.7	NA	NA
1983-85	34.8	11.5	46.3
1988-90	36.1	12.4	48.5
1993-95	36.9	19.5	56.4
1998-00	42.7	25.6	68.3

Source: GVA
Amounts expressed as 3-year averages in million litres of alcohol per year

However, an estimated 30 per cent of the alcohol used by British gin and vodka distillers is not produced from grain but fermented from molasses in Europe and North America (David Ward, Greenwich Distillery; personal communication). Although such fermentations inevitably yield a residual liquor that can be concentrated and marketed as an animal feed, its production outside the country means that the feed is lost to UK livestock producers. Thus, in 1999, the combined production of gin and vodka amounted to 17.6 per cent of scotch whisky production, though the proportion falls to 12.3 per cent when the industries are regarded purely as potential sources of animal feed for the British market.

GNS from British distilleries comprises the remaining 70 per cent of the alcohol used in British gin and vodka, though none of it is produced on-site at any of the gin and vodka distilleries located throughout England from Warrington to Plymouth. Much of the GNS is produced alongside whisky at Scottish grain distilleries, and there appears to be only one site - at Greenwich - where English alcohol is distilled for use in English gin and vodka manufacture. In consequence, none of the gin and vodka distilleries should be regarded as animal feed sources.

In Scotland the production of grain neutral spirit has many operational similarities with grain whisky production, though there are fewer restrictions on the ingredients that may be used. Thus wheat and maize continue to provide the fermentable carbohydrate, but the reliance on malt may be replaced by a combination of exogenous enzymes that includes amylases, ß-glucanases and amyloglucosidases. Mashing proceeds in a similar manner to that employed in whisky production, and again there is the distinction between those distilleries where the wort is filtered and those where both the liquid and solid contents of the mash tun are passed to the fermentation vat. After fermentation, the alcohol is distilled from a continuous still and, in common with those obtained during grain whisky production, the resulting co-products are dried in a dark grains plant or produced and marketed as yeast-enriched moist grains. No attempt is made to separate the co-products of GNS from those of whisky production, and from an animal feed viewpoint this would not seem necessary – except possibly for one point.

GNS is distilled in stainless steel stills in place of the copper stills traditionally used when making whisky. Although a little copper piping may be included to remove sulphurous compounds beyond the still, it is unlikely that the spent wash will contain significant amounts of residual copper, and animal feeds derived from this process may be expected to have a relatively lower copper content. If these co-products were collected and marketed separately, they may prove to be suitable for feeding to sheep as well as cattle. Although this would allow removal of the cautionary note that is always added when most distillery feeds are discussed, it would not in practice represent a significant widening of the market. The hand-feeding of sheep is often confined to a short period around lambing, typically lasting no more than two months. Thus the potential impact of opening up a proportion of this relatively small market on the demand for grain distillers' grains would not be substantial.

At Greenwich and Girvan, the GNS distillery is situated on the adjacent site to the wheat fractionation plant (see Wheat Starch chapter), and uses the residue of gluten and large granule starch extraction as a source of fermentable carbohydrate. At Girvan, the co-product is combined with the residues of grain whisky production and marketed as Vitagold. At Greenwich, there is no solid co-product from this process, but a residual liquor remains after distillation. This liquor is partially evaporated to about 21-23 per cent dry matter and then marketed as a liquid feed to the pig industry under the name Greenwich Gold – more information about this co-product is given in the Wheat Starch chapter.

Co-products of the distillation industry

Draff

Draff is the principal co-product of malt whisky production in Scotland, though much of it is now marketed in the form of a dried blend. Draff, which is derived entirely from the extraction of malted barley, is sometimes referred to as distillers' grains but this description seems vague, since the distillers also produce dark grains, barley, wheat and maize grains and other forms of super grains. The composition and nutrititive value of draff reflect the changes that have occurred during the mashing of the malt. Much of the starch is converted into simple sugars, some of the protein is solubilised, and these and other soluble components are extracted. In consequence, the fibre, oil and protein levels are increased, and so is the gross energy value. Overall, draff represents a substantially different material from the original barley. This distillery co-product is essentially similar to the spent grains that are produced in breweries, but its botanical origin is more restricted and draff typically has less protein, more fibre and a lower digestibility. Draff is widely known in Scotland, where it has probably been fed to cattle and sheep for some five hundred years.

For ruminant livestock, draff is a palatable, moist feed that is relatively rich in protein, and which provides a slow-release energy supply largely contributed by oil, protein and digestible fibre. The major constituent of draff is fibre, and the NDF content is typically 60-65 per cent. During the malting of the barley and subsequent mashing in the distillery, the digestibility of the fibre is increased (Miller 1969), though it remains low. UK Tables report that the digestibility of the organic matter of draff is only 50-54 per cent. Since (in this report) the organic matter contained 21.8 per cent protein (with an apparent digestibility of 74 per cent) and 8.9 per cent oil (at an estimated 90 per cent digestibility), this implies that the remaining components had a digestibility of only 40 per cent. As well as fibre, this fraction includes minor amounts of other highly digestible components, including starch, sugar and lactic acid, and thus the digestibility of the NDF in draff may be expected to be less than 40 per cent. However, because it represents such a high proportion of the feed, the NDF fraction is still likely to provide about half of the digestible organic matter.

Relatively few ME values have been determined *in vivo*; UK Tables report only two widely disparate values, of 9.5 and 10.8 MJ/kg DM (MAFF, 1990), whereas SAC suggests a value of 11.1 for fresh but 10.8 MJ/kg DM for stored draff (Black *et al.*, 1991). Since storage losses are more likely to occur in the digestible fractions, the value for stored draff would seem to confirm the higher value quoted for the freshly delivered material. If the ME value of draff can be predicted from brewers' grains equations developed at the Rowett Research Station (Rowett, 1984b), mean values of 11.6 and 11.3 MJ/kg DM would be suggested by calculation from NDF and NCGD respectively for a limited dataset (James & Son, unpublished information). The E3 ME prediction equation (MAFF, 1993), which was developed for use with compound feeds, but is used more widely in practice for a range of feed materials, underpredicts the ME of the brewers' grains evaluated in the Rowett Study by an average of 14 per cent. Thus its use in the evaluation of draff may be similarly unreliable, and the predicted ME range of 9.6-11.0 MJ/kg DM for the

James & Son dataset may represent underestimates. A notable feature of the digestion of draff is the very low level of methane loss – UK Tables record a loss of only 3 per cent of GE – and this provides some explanation for ME underprediction by the E3 equation. Since lower methane loss equates to a higher ME value, adjustment of the prediction to take account of this factor could add 1 MJ/kg DM – i.e. a 5 per cent saving of the gross energy value of 21.5 MJ/kg DM. This adjustment to the ME prediction from E3 would suggest a range of ME values for the James & Son dataset of 10.6-12.0. Because of the limited information available for such calculations it is suggested that a mean value of 11 MJ/kg DM should be adopted for draff, but the quantitative significance of this co-product would appear to justify further investigation.

It should be noted that 30 per cent of the ME of draff is provided by the oil content, and this significant fraction may be expected to improve the overall energetic efficiency of the animal. This was the consistent finding in the classical studies with dairy cows at Cornell University, where increased milk energy was produced in relation to the calculated fat intake (see Palmquist, 1984). Chudy and Schiemann (1969) have calculated that the direct use of fat in the tissues is 23 per cent more efficient than its synthesis from carbohydrate and, in theory, this additional efficiency could be the equivalent of an extra 0.75 MJ/kg of ME supplied in the form of starch. (Calculation based on 1.23 x 30% of 11 MJ/kg DM). Thus the rich fat content of draff, which reduces energy wastage in the rumen and promotes highly efficient energy use in the tissues, makes draff a feed that is likely to be under-rated by evaluation systems largely designed for other types of feed. The need for further in vivo studies appears to be a logical conclusion.

The loss of available carbohydrate from the original barley results in a co-product feed that can be fed in relatively large amount to both cattle and sheep without concern for its effect on rumen acidity. However, the high oil content of draff may impair fibre digestion if the oil content of the whole diet is excessive, though the addition of additional calcium in the form of limestone may mitigate the effect (Black *et al*, 1991). Providing the oil content of the diet is held within reasonable limits, the high fibre content of draff may be considered as roughage sparing, though it would not normally be used as the sole dietary fibre source. The co-product's roughage sparing role may prove useful in an intensive, high-concentrate diet, as may be used for fast-growing beef cattle and, to a lesser extent, high yielding dairy cows.

During mashing, significant amounts of soluble mineral are extracted from the grain, and this alters the mineral balance of the resulting draff. A report by Scottish Agricultural College advisers (Black *et al.*, 1991) states that draff is virtually devoid of soluble minerals such as sodium and potassium, and is low in some trace minerals and vitamins. When large quantities of draff are fed, it may be advisable to use specially formulated minerals. These should be formulated to ensure an adequate intake of sodium, potassium, calcium, magnesium, selenium and cobalt, but also to make allowance for the characteristically high phosphorus content found in draff.

Draff can be fed fresh from the day it is delivered, or ensiled for feeding some months later. The co-product is naturally acidic and does not require fermentation to ensure good preservation in-silo. However, like all fresh materials, it is vulnerable to aerobic spoilage and even with short-term storage it is important to store the grains carefully and seal effectively. Even with good farming practice, storage will inevitably lead to losses of both quality and quantity, principally as a consequence of effluent and aerobic activity. However, the development of the Grainbeet system, in which draff is ensiled with dry, absorbent sugar beet pulp has made it possible to reduce and potentially to eliminate effluent loss, and to reduce other in-silo losses. Trials at the SAC's Crichton Royal Farm showed a reduction of in-silo dry matter losses from 13 to 6 per cent, when draff was ensiled with 150 kg dried molassed sugar beet pulp per tonne (Hyslop *et al.*, 1989).

In areas close to the distillery, but distant from alternative sources of feed, draff has been fed ad libitum to cattle and fattening lambs. SAC trials recorded cattle growth rates in excess of 1 kg per day on a sole diet of mineralised draff (Black *et al.*, 1991), but such a diet may not make the most of the feed's nutrient supply. Improved performance may be expected when draff is balanced with energy sources such as cereals. When draff is stored together with sugar beet pulp as Grainbeet, the value of the feed is significantly enhanced. SAC dairy trials have fed up to 16.6 kg Grainbeet per day as a concentrate replacer, including approximately 14 kg draff, and achieved similar milk output to that of cows fed a proprietary 18 per cent protein dairy compound. Other trials have shown the value of Grainbeet as a forage replacer, when a 50/50 mix of silage/Grainbeet (containing 6 per cent straw and minerals) was found to stimulate higher intakes, greater milk yield and improved body condition, than a diet containing silage as the sole forage (Perrott, 1993). Bull beef cattle have also performed well on Grainbeet as the sole diet, growing at a rate of 1.6 kg per day. This performance outstripped that of a matched group fed silage and 3 kg per head per day of a barley/soya based concentrate. Store lambs have been fattened on Grainbeet, with a performance level, at 174 g/day, marginally better than that of animals fed on a proprietary compound feed.

In other situations, draff can be fed successfully without beet pulp inclusion. It can form the basis of a buffer feed, mixed with silage or straw and fed to dairy cows at grass. Draff can be used for heifers as an ideal complement to straw or mature hay, and for beef cows; 15 kg draff and 8 kg straw plus vitamins and minerals would provide a suitable diet for suckling dams. In a complete diet for dairy cows, up to 30 per cent of the dietary dry matter has been successfully replaced by mineralised draff; the draff intake amounting to approximately 25 kg per head per day (Hyslop and Roberts, 1989). Sheep can be fussy eaters, but they thrive on draff at all stages - in pregnancy, lactation and as weaned lambs. Sheep on trial at the Rowett Research Station were fed a maintenance diet solely consisting of draff for 28 days (Rowett, 1984a). There were reported to be no refusals but the addition of minerals appeared to increase acceptability.

The importance of mineral supplementation on the nutritional value of draff has also been demonstrated; digestibility has been increased by the addition of either calcium (El Hag and Miller, 1969 and 1972) or magnesium salts (Lewis and Lowman, 1989). The major benefit of these minerals is believed to be due to their reaction with the fat content of draff, and the reduction of any negative effects on fibre digestion in the rumen, rather than to any nutritional benefit of the minerals themselves.

N.B. The solid co-product of some grain distilleries is also referred to as draff and, since this material is different in both origin and nutritional value, there is potential for confusion. However, the solid output from these distilleries does not become available in this form but is blended with spent wash and then dried. Thus the fresh draff that is supplied to farms originates only from malt distilleries.

Pot ale syrup

Pot ale syrup (PAS) is a liquid co-product of malt whisky production, and a concentrated form of the residual liquor that remains in the still after the first distillation of spirit. The unconcentrated pot ale was originally used as animal feed, but the transport cost for such a dilute material became prohibitive. The installation of evaporators in distilleries allowed the production of a much more valuable syrup, and this has been widely used on livestock farms both on its own and as a component of various feed blends. Since the 1960's, a significant proportion of the pot ale syrup has been re-combined with the solid co-product of the distillery and dried into what are commonly referred to as distillers dark grains or malt distillers dark grains.

PAS is a palatable, medium to dark brown, viscous syrup that resembles molasses in physical appearance. It remains fairly homogeneous with only a slight tendency for the solids to separate out (Kennedy, 1987). Pot ale includes a mixture of both suspended and dissolved solids and its viscosity increases rapidly during evaporation. In order to be marketable as a liquid feed, the syrup has to be produced at a much lower dry matter content than molasses. The dry matter content of the PAS produced at different locations may be quite different. A wide range of values of 30-50 per cent is commonly quoted, and Kennedy quoted a range of DM values that stretched up to 60 per cent. Since dry matter content is closely linked to both nutritional and monetary value, PAS has sometimes been sold on a dry matter basis, though this is not universal practice. In nutritional terms, PAS is a high protein material, with a good amino acid balance largely contributed by yeast. The starch and fermentable sugar concentrations are predictably low, but there is a considerable proportion of non-fermentable complex sugars deriving from cell wall components of both barley and yeast (Rowett, 1984c). Since the fermentation process allows the proliferation of lactobacilli, PAS also contains a significant proportion of lactic acid - amounting to some 10 per cent of the dry matter. PAS contains very little fibre but the oil content seems variable, with evidence of *de novo* synthesis by the yeast. Some difference has been recorded between oil values determined with or without acid hydrolysis. There is also a significant ash content, which includes a substantial amount of phosphorus. The presence of phytase enzymes in both malt (Rowett, 1990) and yeast (Anon, 1992b) is claimed to render this phosphorus highly available to non-ruminant animals. This claim is supported by unpublished work at the Rowett Research Institute (1990) where it was found that only 5.1 per cent of the phosphorus in a sample of PAS was present as phytate phosphorus.

PAS is reported to contain a factor that stimulates rumen micro-organisms (Topps, 1983), and particularly the cellulolytic bacteria. This factor, which is believed to be associated with the yeast fraction, appears to stimulate the intake and digestibility of roughage feeds. Gizzi (2001) suggests that there may also be a stimulation of lactic acid utilisation, which would be helpful with high concentrate diets.

The digestibility of PAS by ruminant animals is high, and the organic matter digestibility has been measured in two trials at 89 and 93 per cent (Rowett, 1984a and c). The PAS used in these evaluations had a high gross energy content of 20-20.4 MJ/kg DM, and thus the high digestibility resulted in high DE and ME values. The mean ME value was determined at 15.6 MJ/kg oven dry matter. SAC advisers have recommended a more conservative ME value of 14.2 (Black *et al.*, 1991), but no *in vivo* data have been published to support this value. There may be some concern that the GE value determined in these trials is higher than the composition would lead one to expect, but the Rowett report highlights the fact that their analysis fails to account for approximately one-third of the dry matter. In this uncertain situation, it would seem reasonable to accept the Rowett measurements as the best available facts.

The crude protein content of PAS, at 34-38 per cent of the dry matter, confirms this as a protein-rich feed. However, in the Rowett studies, the digestibility of the protein, at a mean value of 81.5 per cent, was a little lower than the rest of the organic matter. This may have been a consequence of exposure to high temperature during the evaporation process, and it may thus be notable that evaporation at lower temperature is now carried out at some facilities. SAC notes that much of the crude protein consists of simple forms such as peptides and amino acids and, consequently, "the protein is virtually all degradable in the rumen" (Black *et al.*, 1991).

PAS has a low pH value in the region of 3.5 and is usually regarded as having good keeping qualities. This is an important consideration because the demand for the syrup as a ruminant feed tends to be seasonal. However, some PAS samples have been found to be unstable, leading to microbial degradation and associated gas production on storage (Johnstone, 1991). Further investigation revealed that the less stable materials had been produced by a low temperature evaporation process, which may have failed to sterilise the co-product. In a response to this potential problem, propionic acid is added as a preservative to the PAS at some distilleries – though this by no means universal practice.

The definitive feature of a malt distillery is the copper still, and the corresponding characteristic of pot ale is its high copper content. A wide range of copper level has been reported, with values commonly in excess of 100 mg/kg DM. Such a level may constitute a potential risk if PAS were fed to sheep, since some breeds are known to be extremely intolerant of high dietary copper levels (Suttle, 1999). However, the copper content of PAS has been shown to be poorly available; in a copper repletion study, the availability of PAS copper was found to be only 1.1 per cent - less than one-fifth of the availability of copper supplied as copper sulphate (Suttle *et al.*, 1996). This may explain why PAS has been used without apparent problem by some sheep farmers. But it would be prudent to blend it with other low-copper feeds and to avoid using PAS at a high inclusion rate, or for prolonged periods, or to breeds that are known to be particularly susceptible.

Cattle are more tolerant of high copper levels in the diet and PAS may be fed to them in a variety of ways. It has been poured onto coarse roughages such as hay or straw, self-fed through ball and lick feeders or blended with molasses and condensed molasses solubles and offered in the same way. Probably the most effective feeding method is to incorporate PAS into complete diets, where its high energy and protein levels can be balanced most effectively. As a low pH value liquid feed the acidity of PAS needs to be neutralised, but its liquid form does not stimulate salivation and the addition of alkali to the rumen via this route. Thus, the combination of PAS with roughage feeding may be seen to be beneficial.

PAS has also been fed to pigs, and work at the Rowett Research Institute (Rowett, 1983 and 1985) has shown similar performance in growing pigs where the syrup replaced up to 30 per cent of a barley/soyabean diet. A proportion of the soluble component is fermented in the hind gut, and the digestion products of this process are likely to be used less efficiently than those produced by digestion in the fore-gut. SAC advisers (Black *et al.,* 1991) have "corrected" the estimated DE value to take account of the less efficient use of the products of hind-gut fermentation, but such an adjustment is open to argument. Digestible energy is a defined term which makes no allowance for the efficiency of utilisation of digested components, and thus "corrected" values are wrong by definition. They may also be difficult to use in practice when comparison is being made with the uncorrected DE values given for other feeds.

The lysine content of PAS was measured by Rowett workers (Rowett, 1990) to be 6.35 per cent of the protein. This is in line with the lysine content of the protein in soyabean meal and brewers' yeast (6.2 and 6.6 per cent of protein respectively). Other essential amino acid levels are also comparable with those of soyabean, indicating that PAS is potentially a good quality protein source. However, Black *et al.* (1991) claim that PAS protein is of poor quality as a result of the "severe heat treatment during its production". This damning comment is later qualified by reference to some syrups now being produced by a low temperature process. The PAS examined in detail in the Rowett (1990) study was reported to have a lysine digestibility measured in the chicken of 73.5 ± 6.7 per cent, which appears to be in line with many feeds from other sources. Whilst heat damaged protein is likely to be reflected in poor amino acid availability, it seems reasonable to assume, given the industry's adoption of low temperature systems, that this is not the invariable result of pot ale evaporation.

Barley distillers dark grains / malt distillers dark grains

Malt distillers dark grains (MDDG) are the principal form in which the co-products of malt distilleries are marketed, and this predominance is expected to continue (Pass and Lambart, 2001). They are the dried form of a blend of draff and pot ale syrup and are widely used as a ruminant feed throughout Britain. Both components represent material from which much of the fermentable carbohydrate has been extracted, although the pot ale fraction contains a significant proportion of more complex sugars. Whilst the draff is derived exclusively from malted barley, the pot ale comprises a mixture of residues from both barley and yeast. Most MDDG drying plants now dry the co-products from a number of distilleries, and the mixing of material from different sources may be expected to reduce previous variation in the proportions of the blend, and in the composition of the final material.

Barley distillers dark grains can be fed locally or transferred to cattle feeding areas across the English border.

MDDG are a feed with high protein and high gross energy contents. Typical protein levels would be in the range of 23-29 per cent of the dry matter, whilst gross energy values of 20.9-21.9 MJ/kg DM have been reported (Gizzi, 2001; MAFF, 1990). However, MDDG are a fibrous feed, with an NDF content in excess of 40 per cent, and for this reason the material would not usually be recommended as a pig feed. For ruminants, the digestibility of the feed is a crucial determinant of its nutritional value. Gizzi (2001) quoted a relatively narrow range of organic matter digestibility values of 64.8-68.5 per cent from four in vivo trials, while two energy digestibility measurements – both of 67 per cent - are reported by MAFF (1990). The methane loss during the digestion of draff is notably low and MDDG reflect this, though to a lesser degree, with an average methane loss of 4.7 per cent of GE. The determined ME value from five studies was reported by Gizzi to lie between 12.1 and 13 MJ/kg DM, with one anomalous value measured at 13.9 MJ/kg DM.

In common with other distillery feeds, the ME value of MDDG cannot be predicted with acceptable accuracy from widely used equations, such as E3 (Pass *et al.*, 1989). Wainman *et al.* (Rowett 1984d) suggested two alternative equations but both included parameters – acid detergent lignin and Christian lignin - that are not routinely determined. Gizzi (2001) suggested an equation to predict the ME value of barley, wheat and maize distillers dark grains based on GE, NDF and the ME of the relevant cereal. The latter term is presumably intended to be a constant, not an invitation to consider the wide range of ME values that have been determined for cereal grain but, since GE is not commonly measured, this equation is no more utilisable than that of Wainman *et al.* There would thus seem to be some justification for practical purposes in adopting the mean value of 12.6 MJ/kg DM suggested by the six studies reported by Gizzi.

A key component of draff is the high oil content, which may be as high as 13 per cent of the dry matter. Such an oil content may be expected to improve the efficiency with which ME is utilised, but caution would be needed when feeding in order to avoid the effect of an excessive oil level in the overall diet on fibre digestibility. With MDDG, the oil content is moderated by blending with pot ale syrup, and levels reported by Gizzi to lie between 6.5 and 9.5 per cent could be fed in a well-balanced ration without undue concern. Conversely, the mineral content of MDDG is higher than that of draff - much of the mineral extracted in the mash tun is returned when the pot ale syrup is re-combined. This still leaves a feed that has only a low calcium content, but the phosphorus content represents a rich supply. High copper is an inevitable feaure of a blend that includes pot ale syrup, though the inclusion of draff dilutes the copper content of this feed. The poor availability of the copper in PAS is reflected in MDDG; an availability of 3.5 per cent was measured by Suttle and colleagues (1996), which was intermediate between PAS and an inorganic copper supply. This implies that MDDG may be used for cattle and, with care, for non-susceptible breeds of sheep – though it should not be fed to indoor sheep or for prolonged periods.

The value of the protein supplied by MDDG has given rise to considerable debate. The extraction of the soluble protein from malt during mashing leaves the draff with a relatively insoluble protein content, but the addition of pot ale syrup in the manufacture of MDDG, brings a rich source of soluble protein. After the mixing of these two components the moist blend is dried, and that heating may result in some depression of protein digestibility (Mitchell, 1990). However, some of the soluble protein, and some of the fine particles previously regarded as degradable protein, are believed to escape digestion in the rumen (Offer, 1992). Moreover, some of the insoluble protein that passes through the rumen is considered to be more digestible than previously thought. For most feeds, the ADIN fraction of the crude protein is regarded as indigestible (Webster and Chaudhry, 1992; AFRC, 1993) but calculations made by van Soest and Mason (1991) showed that 58 per cent of the ADIN in distillers' grains was digestible, and this indication is supported by in vivo studies with rats (Whyte, 1993). While this debate continues, an annual volume of 60-100,000 tonnes is being consumed with no report of poor performance due to low protein availability (Pass and Lambart, 2001). This satisfactory impression is supported by published research; 7 kg of MDDG fed in place of a similar amount of a conventional compound feed supported marginally higher outputs of milk and milk protein, though there was a reduction in the fat content of the milk (McKendrick and Hyslop, 1992).

Dried dark distillers grains can be fed in the milking parlour

(courtesy: Guinness UDV)

Grain distillers grains – wheat

Grain distillers grains (GDG) are the principal co-product of grain whisky and grain neutral spirit production, and the sole feed material available from five of the eight grain distilleries in Scotland. Their choice of cereal grain has been dictated by the relative prices, and for many years the industry was based on maize, but wheat is now the predominant raw material. Grain distillers grains-wheat (GDG-W) are a dried blend of two co-products from the distillery: the cooked and extracted wheat grains left after mashing – with a minor inclusion of extracted malt – and an evaporated form of the spent wash (ESW) that remains in the still after the removal of the alcohol. GDG-W are the grain industry's equivalent to the malt distillers dark grains, but their nutritive value is significantly higher.

The protein content of GDG-W is high, with a typical content of 30-35 per cent. These values are higher than those commonly found in malt distillers grains (23-29 per cent), and would seem to reflect a smaller proportion of extracted grains and a greater proportion of the yeast-rich spent wash. However, such a variation in the proportions would be expected to affect the amino acid composition of the GDG-W protein, but there appears to be no confirmation of that. Lysine values of 0.50-0.77 per cent of the dry matter are reported in UK Tables (MAFF, 1990) for 5 samples of GDG-W, and these are much more in line with wheat protein than with yeast.

The fibre content of GDG-W is quite variable, and UK Tables include a range of NDF values of 23-46 per cent of the dry matter. However, the crude fibre fraction is much more consistent, with a range of only 6.8-10.4 per cent for the same 10 samples. This apparent discrepancy may be partly due to analytical difficulty with the NDF determination – such difficulty has been experienced by the author when analysing a wheat syrup. The digestibility of the organic matter is reasonably consistent, seven values identified by Gizzi (2001) all lay within the range 70-77 per cent. A mean OMD value of 74 per cent is distinctly higher than the 65 per cent measured in MDDG. This may be due to the innately higher digestibility of the fibre in wheat as opposed to that in barley (67 v 46 per cent on average, MAFF, 1990), or to an increase in the digestibility of wheat fibre during pre-cooking.

An average oil content of 5-6 per cent of the dry matter, 6-7 per cent after acid hydrolysis, would seem to reflect a dilution of the relatively high values that are commonly found in extracted grains. However, oil contents of up to 9 per cent have been found on occasions. This fraction may be regarded as providing a useful energy source for ruminants, without need for undue concern about any negative effects on fibre digestion.

The gross energy value of GDG-W is consistently high; Gizzi (2001) reported values from a number of sources that all lay in the range 21.0-22.6 MJ/kg DM.

Taken together with a mean energy digestibility of 73 per cent (MAFF, 1990), this establishes a probable DE value in the range 15.3-16.5 MJ/kg DM. Methane loss is reported by ADAS (1988) to average only 5 per cent of GE and, together with a urinary loss of 7 per cent of GE, this leads to an expected ME value of 12.7-13.9 MJ/kg DM. While this range suggests a typical ME value higher than the 12.2-12.8 reported in UK Tables, it is lower than the value of 14.5 MJ/kg DM suggested in commercial literature (Trident I).

Copper stills play a traditional role in grain whisky production as they do in malt distilleries, and the copper content of the residual spent wash is predictably high. Where grain neutral spirit is produced on the same site, the spent wash is likely to have a much lower copper content, but the co-product of GNS production forms only a minor proportion of the liquid component used in the manufacture of GDG-W. The blending of the spent wash with the extracted grain residues may be expected to dilute the copper level of GDG-W. However, a wide range of copper level of 14-140 mg/kg DM was found in 5 samples tested by ADAS (1988). While the highest value in this study may be considered to be anomalous, a recent report (Gizzi, 2001) noted that recent modifications to the grain distillation process had "markedly increased the copper content of DDG from some Scottish grain distilleries - up to 80 mg/kg DM". Thus there must be continuing concern about the risk of copper toxicity, and particularly for copper-sensitive breeds of sheep. The availability of the copper in GDG-W was tested recently, in comparison with other distillery feeds and an inorganic copper sulphate standard (Suttle *et al.*, 1996), and the risk was confirmed. A copper availability of 6.4 per cent in GDG-W was shown to be higher than that from copper sulphate (5.7 per cent).

Thus, the copper content of GDG-W makes the feed inadvisable for sheep, but it represents a very palatable feed for cattle. Satisfactory performance by dairy cows was obtained when McKendrick and Hyslop (1992) fed 7 kg GDG-W in place of an equivalent amount of compound feed.

Grain distillers grains – maize

As noted in the previous section, grain distillers grains (GDG) are the principal co-product of grain whisky and grain neutral spirit production in Scotland, and they are produced in a similar manner from either wheat or maize. In the manufacture of GDG-M, the solid grains left after the cooked maize has been mashed are re-combined with the evaporated spent wash from the stills, and the mixture is then dried.

GDG-M have a typical protein range of 27-32 per cent of the dry matter, which is a little lower than the protein level of GDG-W but higher than that of MDDG, the comparable barley product from the malt distilling industry. However, the oil content of GDG-M, at 9-13 per cent of the dry matter is significantly higher than the oil content of other types of dark distillers grains. This is a direct consequence of the much higher oil content of the maize grain, and the oil content is further concentrated by the extraction of available carbohydrate during mashing.

The extraction process has a dramatic but apparently variable effect on the fibre content of the co-product, lifting the NDF to a wide range of values. UK Tables quote NDF values of 23-41 per cent of the dry matter, but Gizzi (2001) lists two NDF values from other studies of over 50 per cent. Although some of this variation may be due to analytical difficulties, as noted in the discussion of GDG-W, there is notable variation even in the lignin content of this feed material. UK Tables show a range of lignin content of 5.0-13.4 per cent in a relatively small survey of five samples.

The organic matter digestibility of GDG-M has also proved to be variable, with a range of values of 66.5-84.0 per cent from only six tests carried out by the RRI (Rowett, 1984d) and ADAS (1988). The mean value of 75 per cent is marginally higher than the mean value calculated for GDG-W. However, the higher oil content of the maize co-product enhances the GE content; eight GDG-M gross energy values summarised by Gizzi (2001) show a mean value of 22.3MJ.kg DM from a range of 21.7-23.65. The high oil/low starch composition of GDG-M is also associated with low methane losses in the rumen, and values of 1.9-6.0 per cent of GE have been recorded. The combined effect of these factors is to produce a range of ME values – five of the six values measured by RRI and ADAS were in the range 13.6-15.4 MJ/kg DM, with one apparently anomalous value of 17.3. For routine ration formulation purposes, it would seem reasonable to use the mean of this range; viz: 14.9 MJ/kg DM. For other purposes it would seem desirable to develop an ME prediction system that recognises some of the innate variability of this feed.

As with GDG-W, the copper stills impart a high copper loading on the spent wash and render this feed unsuitable for feeding to sheep, but GDG-M has been used with some success for feeding to cattle. The co-product was used by Owen and Larson (1991) for cows in early lactation, and milk production exceeded that of cows fed a diet based on soyabean meal, when GDG-M constituted 19 per cent of the diet though not when fed at a 34 per cent inclusion rate. The possibility that GDG-M is best fed in restricted quantity – or in better balanced rations – was also indicated by the results of McKendrick and Hyslop (1992). They found that, at 7 kgGDG-M per head per day, milk yield, protein and fat were all reduced in comparison with a similar amount of a conventional compound feed.

Supergrains™

Supergrains are the moist co-product of grain whisky and neutral grain spirit production at Cameronbridge, and they constitute the only feed co-product from this site. The name is unique to the production from this distillery, though the material has also been referred to as wheat distillers centrifuge cake. In former times, the principal raw material used by the distillery was maize, but wheat is now established as the preferred cereal. The grain is cooked, mashed and fermented before distillation, and the spent wash that remains is the basis for the Supergrains output. The spent wash is transferred into large decanter centrifuges, where the solids are separated from the liquid effluent. At this point where Supergrains leave the processing, at a temperature of 75-80°C, they are relatively sterile. But experience has shown that they that they are potentially vulnerable to aerobic decay, and before Supergrains are consigned to the animal feed silo they are treated with propionic acid as a preservative.

Supergrains comprise the non-starchy components of the wheat grain, plus the yeasts that multiplied vigorously during fermentation. Supergrains can largely be attributed to crude protein (28-32 per cent), oil (10-11 per cent) and fibre (NDF 60 per cent), with small amounts of starch, sugar and ash. Although these figures provide a realistic and repeatable description of the co-product, it is not uncommon for the analytical fractions to add up to to more than 100 per cent. However, Supergrains are clearly a rich source of protein, and they have proved to be a substantial energy source too.

A number of GE values have been reported between 22.0 and 23.7, with a mean value of 22.8 MJ/kg DM. These are a little higher than the GE values reported for the dried co-product produced in other wheat distilleries, which include the soluble components, and the discrepancy is consistent with the higher oil content found in Supergrains. Energy digestibility measurements have been relatively consistent, with a mean value of 72.1 per cent, which is close to the value adopted for GDG-W. The combination of GE and digestibility data suggests that Supergrains have a measured DE value of 15.9-17.1, with a mean value of 16.4 MJ/kg DM. No measurements of methane loss appear to have been made and authors have made use of conversion factors to calculate the ME value. If one accepts that the urinary energy loss will be similar to that measured for GDG-W (i.e. 7 per cent of GE), and that the methane loss is likely to be proportional to the oil content, the latter may be estimated and used to calculate ME. The author's estimate is based on interpolation of methane measurements made at the RRI on a series of draff and barley meal mixtures (Wainman, Dewey and Brewer, 1985). At 80 per cent draff: 20 per cent barley meal, the oil content may be similar to the typical value measured in Supergrains, and the estimated methane value would be 3.3 per cent of GE. Thus, the combined loss of urinary and methane energy is estimated to be 10.3 per cent of GE – i.e. equivalent to 2.3 MJ at the mean GE value. Deduction of this amount from the DE range suggested above, leaves an ME range of 13.6-14.8, with a mean value of 14.1 MJ/kg DM. This value is significantly higher than that derived for the dried product (GDG-W), though this may be explained by the higher oil content of Supergrains.

In common with other distillery feeds, ME prediction by commonly-used equations is unlikely to produce values in line with those determined by in vivo studies. Pass and co-workers (1989) particularly noted that the E3 equation (MAFF, 1993) underpredicted the ME value of dark distillers grains by a considerable margin. Two alternative equations were suggested by Pass *et al.,* based on GE and crude fibre, but they appear to be inapplicable to Supergrains. A further equation suggested by Gizzi (2001) for dark distillers grains, based on GE and cereal ME, also appears to underpredict the Supergrains ME value. Thus, for routine ration formulation purposes, adoption of the mean in-vivo value would appear to be preferable.

Supergrains have been fed in measured studies to beef and dairy cattle. In the beef trial of Lewis and Scott (1990), Supergrains and minerals incrementally replaced 4.5 kg of a barley/soya/molasses based compound feed. Growth rates of 1 kg per head per day were targeted and, in practice, slightly exceeded by all groups. At the higher inclusion rates of Supergrains (9.3 and 14 kg per head per day), the condition score of the animals was considered to be poorer and perhaps suggesting a need to finish the cattle at a heavier weight. It seems probable that the high protein, low starch levels of Supergrains may be unsuitable as the sole supplement to a silage-based diet for finishing cattle, and that a balanced diet with cereal feeds could be used to better effect.

Dairy cows were fed 0, 9.6 and 19.2 kg Supergrains together with either barley or a compound feed in a trial by McKendrick and Hyslop (1991). The milk yields were broadly similar, though Supergrains were associated with a reduction in milk fat production – by 2.8 and 7.9 per cent compared with the compound feed control diet. This finding was in line with the result of feeding Grainbeet (draff and sugar beet pulp) in a previous study, and seems likely to be associated with an effect of fat on ruminal digestion. However, the fat content of the diets was only 4.6 and 5.5 per cent of the dry matter, and these levels are much lower than the critical fat levels identified in a survey of the literature by Wu and Huber (1994).

Careful attention to mineral supplementation would seem to be necessary. Supergrains have relatively low levels of all mineral elements, and the additional importance of calcium and magnesium in ameliorating the effects of the oil content has been noted in studies with draff (El Hag and Miller, 1969 and 1972; Lewis and Lowman, 1989). Copper levels have been found to vary widely (29-92 mg/kg DM; James & Son, unpublished data) and Supergrains would not generally be considered suitable for feeding to sheep.

Vitagold™

Vitagold is the moist feed produced at the Girvan grain distillery and is unique to that operation. It is the only co-product feed produced by the distillery, and it contains virtually all the suspended solids that remain in the spent wash after the alcohol has been removed by distillation. This solid material is separated by filter press from the thin effluent, and the press leaves a high value feed with a dry matter content of approximately 34 per cent. Vitagold consists of the extracted residues of wheat and a small amount of malt, together with yeast debris that remains at the end of the fermentation. The wheat that makes up the bulk of the raw material is pressure cooked before mashing and this not only facilitates the conversion of starch to sugar, it also improves the digestibility of the fibre.

The extraction of available carbohydrate from the grain leaves a residue that is richer in protein, oil and fibre, and the protein is further enhanced by the yeasts that proliferated during the fermentation stage. Although information appears to be lacking, it seems likely that - in common with some other distillery co-products - Vitagold will also have a high GE content. Vitagold typically has a protein content of 36-38 per cent, an oil content of 8-12 per cent and an NDF level of approximately 50 per cent of the dry matter. These fractions account for most of the dry matter, together with a small ash content. Since specific in vivo information is lacking, the energy values and energy losses during digestion have been assumed to be in line with those for Supergrains.

The relatively high oil and low ash content implies that adequate mineralisation will be needed to ensure adequate intakes of calcium and magnesium. Additional quantities of these mineral elements may also be expected to reduce any deleterious effect of the oil on fibre digestibility. Trace element supplements would also be needed, though they need not include copper. In common with so many distillery feeds, Vitagold offers a rich copper supply – sufficiently so, that it would not normally be recommended for sheep feeding.

Loch Lomond Supers™

Loch Lomond Supers (LLS) are a moist, solid co-product of grain whisky and neutral spirit production at the Loch Lomond Distillery. In the initial stages, the wheat grain is ground and then cooked before mashing with a small proportion of malt. Beyond the mash tun, the whole mass of solid grains and liquid extract is carried through the fermentation and distilling processes. On completion of these stages, the spent wash is centrifuged to produce two fractions, both of which are used as animal feed. LLS constitute the solid portion, while the liquor – together with a small proportion of pot ale from malt distillation on the same site – is evaporated and sold as a syrup under the name Loch Lomond Gold.

The production of LLS by centrifuging the spent wash results in a co-product that has similarities with the Supergrains from Cameronbridge. However, the use of the centrifuged liquor as a feed, as opposed to its disposal as an effluent, gives Loch Lomond greater flexibility in its decanter settings. Whereas the Cameronbridge objective is to remove the maximum amount of suspended solids from the spent wash, at Loch Lomond a proportion of the solids can be switched between the two fractions. This typically allows the LLS to be marketed at a slightly higher dry matter content, though the corollary may be greater viscosity of the syrup fraction.

LLS are rich in protein and high in fibre and oil. Compared to Supergrains, they have a lower NDF content (48 v 60 per cent of the dry matter) and a slightly higher digestibility, as measured by the NCGD laboratory procedure. However, the oil content at 8 per cent of the dry matter, is lower than Supergrains and more in line with the oil content of wheat dark distillers grains. A further notable difference is the relatively low copper content of LLS, which is a reflection of the limited use of copper in the Loch Lomond grain distillery. This co-product could be used as a feed for both cattle and sheep without undue concern for its effect on breeds of sheep that may be susceptible to copper toxicity.

There have been no animal evaluation studies with LLG, and thus the ME value has been calculated from Supergrains data. The suggested ME value of 13.6 MJ/kg DM is lower than that determined for Supergrains, but marginally higher than the ME calculated for dried dark distillers grains of wheat origin. This latter difference may appear inconsistent; GDG-W are of similar oil content and may typically be expected to have a lower fibre. There may be a partial explanation in terms of a depression of digestibility as a result of drying, as noted by Mitchell (1990), but in vivo confirmation would be helpful.

Loch Lomond Gold and evaporated spent wash

Loch Lomond Gold (LLG) is the liquid feed co-product of whisky and grain neutral spirit (GNS) production at the Loch Lomond distillery. It largely represents the liquid fraction of the spent wash that is separated by centrifuge from the extracted wheat grains produced in the grain distillery, but also includes a proportion of pot ale syrup from malt whisky production on the same site. A blend of these two liquids is evaporated to yield a mid to dark brown syrup that is used as a liquid feed for both cattle and pigs.

LLG has a typical dry matter content of 30-32 per cent, which is somewhat lower than the DM level at which the co-product was previously marketed. It is also lower than the characteristic dry matter content of pot ale syrup. The viscosity of LLG tends to increase more significantly than that of pot ale syrup when its dry matter content rises above 35 per cent owing to a higher content of non-starch polysaccharides (NSP). NSP occur in the endosperm cell walls of barley and wheat in long molecular chains that have a tendency to aggregate as water is lost from the evaporator, causing an increase in viscosity in the residual solution. Although barley, from which the pot ale derives, has a higher content of ß-glucans than wheat, these long chains are largely broken down by the enzymes produced during malting, and aggregation is less of a problem. The ß-glucans in wheat, together with other NSP compounds, are largely undegraded during the various stages of grain whisky and GNS production, and they remain as a problem in the spent wash. ß-glucanase enzymes have been used at Loch Lomond in attempt to reduce the problem and they may be used again. But enzyme treatment adds to the cost, and its economic justification depends on the extent of the problem – which varies from year to year – and the likelihood of its recoupment through higher dry matter, higher price and reduced haulage costs.

LLG is a useful source of both energy and protein for ruminants and pigs. At 25 per cent of the dry matter, the protein content of the syrup is lower than that of the corresponding grains, but it represents twice the protein content of the original cereal and LLG provides a valuable source of rumen degradable protein. Although some of this protein derives from the yeasts that multiplied during fermentation, the amino acid profile more closely resembles that of the cereal. LLG is not an ideal source of protein for the fast-growing pig.

No in vivo assessments have been made of the energy value of LLG, and no gross energy value has been reported. Thus the calculated energy value is somewhat speculative. An estimated GE value has been calculated from the analysis and a digestibility of 90 per cent assumed in line with the measured digestibility of pot ale syrup, but lower than the value suggested by the NCGD analysis. Urinary and methane losses in line with those for pot ale syrup have been assumed, and these assumptions suggest an ME value for ruminants of 14 MJ/kg DM. If one also assumes a digestibility of 90 per cent by the pig, this would imply a DE value for pigs of 16.8 MJ/kg DM. The value of that DE to the pig may be affected by the proportion that is fermented and absorbed from the hind-gut, but this cannot be quantified without further study. Since the efficiency of DE utilisation is outside the classical definition of DE, no adjustment of the calculated figure has been made. A DE value of 16.8 MJ/kg DM is thus suggested but the need for in vivo assessment is emphasised.

LLG is regarded as a palatable ration ingredient that mixes readily with other components, imparts its own characteristic flavour, reduces dust levels and prevents the separation of costly micro-ingredients. It has only a low calcium content, but is a rich source of phosphorus. Its sodium content is low, but this is accompanied by a significant potassium level. The copper content of LLG is moderately high but, since the grain distillery carries a minimum of copper pipework, much of this would appear to derive from the pot ale syrup. Given the poor bio-availability of copper from this source (Suttle *et al.*, 1996), and its dilution in LLG, this material would seem to pose only a limited risk of copper toxicity for most breeds of sheep.

LLG has been fed to both growing pigs and sows and an intake of 15-20 per cent appears to have been satisfactory. For ruminant stock, LLG has typically been used as a molasses replacement and offered similar virtues, in terms of good palatability and an effective mineral carrier. It would also appear to offer better nutritional value – more energy and protein, and particularly for those on a low plane of nutrition more phosphorus, magnesium and trace elements. Adequate fibre sources would need to be used in conjunction with LLG but, at 5-8 kg per head per day this co-product would make a significant contribution to the nutrient needs of beef and dairy cattle.

Evaporated spent wash: A proportion of this feed is available in liquid form from some grain distilleries where the majority of the co-product is blended and dried as dark distillers grains. Its dry matter content is variable though, because of viscosity problems, it tends to be lower than that of pot ale syrup. In this , and in other respects, this feed is similar to Loch Lomond Gold – which itself is largely a blend of ESW with some pot ale syrup.

7 Maize Fractionation

The processing of maize for the production of starch and syrups is a huge industry in the United States that consumes some 30 per cent of the US maize crop (M Hardy, Cargill PLC; personal communication). The co-products of these operations are also produced and exported on a huge scale, and they have been for many years - an export volume of one million tonnes of maize co-product from the United States was exceeded for the first time in 1942 (CIRF, 1959). Maize gluten feed, in particular, has long been established on the European animal feed market.

In Britain, the fractionation of maize into its constituent parts was first practised in the textile industry, where Messrs. Brown and Polson patented a procedure in 1852 that enabled them to use the extracted maize starch to stiffen cloth. More than a hundred years later the Brown and Polson business was purchased by Cerestar, and this company remains one of only two involved in maize fractionation at the present time in Britain – the other company being Cargill plc, which is also a major player on the American scene. Despite the small number of British maize processors, and the fact that production is totally based on imported raw material, the scale of the industry has grown in importance - by the year 2000, some 750,000 tonnes of maize per year were being processed in Britain.

The fractionation of maize was reported by Wainman and Dewey (1988) to result in the following approximate proportions: 61 per cent starch, 20 per cent animal feed, 4 per cent oil and 16 per cent water. If these proportions are applied to the current British processing volume, they imply an animal feed component of 150,000 tonnes of maize co-product dry matter. This may be an underestimate - Firkins *et al.* (1985) suggested that the major feed co-product, corn gluten feed, itself contained 28 per cent of the original weight of the corn. Apart from the broken grain separated at the start of the process, all of the maize fractions are produced in moist form, though only a small proportion of the co-product feed materials are marketed as such. The majority are dried and used as ingredients by the compound feed industry or as straight feeds by livestock farmers. However, this has not always been the case; during the 1980's there was increasing concern about drying costs and, in the United States, there was much debate about a possible change in the duty-free status of feeds from maize processing that were exported to the EC.

These concerns stimulated a flurry of research into the nutritional properties of moist maize co-products. For a time, moist maize gluten feed was marketed as an animal feed in Britain and it is expected that renewed supplies will become available from the Autumn of 2001.

Maize can be fed as the whole grain or as processed fractions

Genetic manipulation

Having continued its development over a 150 year period, the industry had to face one of its greatest challenges at the end of the twentieth century. An increasing proportion of the American crop was being grown as new varieties, which had been genetically manipulated to confer pest resistance - to the corn borer pest in particular - with a consequent benefit in terms of reduced pesticide use (Thomas, 2000). But this expansion came to a crashing halt when the British and later the European public set its collective face against what was perceived as technological witchcraft. Public pressure, expressed through the enormously powerful buying departments of supermarket chains, dictated that no foods containing genetically modified (GM) materials should be offered without appropriate labelling. In practice, the insistence on GM labelling, with the implication that consumers would be offered a choice, quickly developed into an effective ban on GM foods. For the maize starch industry, whose products had found their way into a wide range of food items, it rapidly became necessary to source its maize with great care.

All of the UK raw material is now derived from South-West France, where vast quantities of non-GM maize varieties are grown and stored in dedicated silos. GM varieties, by comparison, are grown only in isolated pockets. Because of the vagaries of cross-pollination, even this careful choice of origin cannot guarantee GM-freedom and, prior to the maize being shipped, samples are checked in an independent laboratory by Polymerize Chain Reaction (PCR) analysis. Today, the whole purchasing procedure is required to conform with a protocol so that it can be proved that maize entering the processing factory fulfils the following requirements:-

- The maize was produced from non-GM seeds.
- All maize products and by-products can be traced back to the boat on which the raw material arrived, and harvested crops can be traced forwards to the boat on which they were exported.
- Due diligence has been taken to minimise the adventitious presence of GM maize from field to factory.
- Maize from GM varieties is isolated under all circumstances.

Even with all these measures it is still impossible to give a guarantee of complete GM freedom, but it is generally recognised that, when every precaution has been taken at all the steps along the supply route, and this definition of GM freedom appears acceptable to most people. However, whatever the risk presented by the consumption of GM-modified maize, it is considerable lessened by the processing of the grain. The fragmentation of DNA, which occurs during processing, greatly reduces the possibility of functional genes, either native or modified, being present in the resulting feed material (Deaville *et al.*, 2000).

Quality assurance

The decision to source maize from Southern France is also influenced by the processing industry's preference for the softer dent corn. Further north, only the earlier-maturing flint corn is grown, and this has been found to be more abrasive on machinery and to lead after processing to a lower starch yield. Maize procurement is documented in ISO9002 procedures and, in addition to GM checks, auditors investigate other quality parameters including temperature, moisture, the presence of impurities, broken grain and evidence of insects. Stringent standards for the presence of aflatoxin in animal feeds make this a good point to carry out the necessary checks, or elicit satisfactory reassurance prior to purchase.

Well-defined cleaning procedures are in place for all the equipment that will be used along the harvest chain - involving grain trailers, augers, trucks, platforms, silos and driers – and some of these are duplicated at the port of dispatch. A register of approved transport is held at the port, and both silos and freight filling equipment are dedicated to the handling of non-GM materials only.

On arrival at the British factory, the grain is handled in enclosed unloading facilities and again screened on belt conveyors. Any dust that has been collected during shipping will either fall through the 1.5mm screens or be aspirated away. Such wastes are disposed of in an appropriate manner and are not permitted to re-enter the feed chain. Any iron-containing contaminants, which may have a potentially devastating effect on the processing machinery, are also removed at this initial stage by electro-magnets. The cleaning operation is considerably aided by the large size of the maize grain although, after multiple handling during initial storage, shipping, offloading and internal transfer, broken grains constitute an inevitable fraction of any consignment. Such broken material is much less suitable for processing and this fraction is diverted without further attention to the animal feed market. For livestock farmers, who would need to roll or hammer-mill whole maize grain before use, **broken maize** represents a high energy, ready-to-use feed material.

Before the screened maize is processed, it is subjected to a number of other quality checks. Both broad and full screening tests are regularly carried out for the presence of mycotoxins, and chemical assessments are made on a routine basis. Rapid answers are required at this stage, if the testing is not to hold up the movement of raw material into the factory.

Wet milling

The isolation of maize starch initially proved more difficult than the extraction of starch from wheat, potato or cassava. Although maize grain contains a high proportion of starch (about 70 per cent on a dry matter basis), its separation is impeded by an unusually high concentration of oil (cf maize 4-5 per cent, wheat 1.5-2.5 per cent, roots and tubers 0.2-0.4 per cent oil on a dry matter basis). The problem was overcome by the development of a wet milling procedure (CIRF, 1959), in which the oil-rich germ was separated before the grain was ground to facilitate further fractionation.

Maize fractionation

The initial part of processing comprises a steeping of the grain in slightly acidified water at about 50°C. Steeping is a dynamic, counter-current leaching process that proceeds over a period of 30-40 hours, as the maize advances through a battery of steeping vessels, with the composition of the steeping liquor changing progressively. This process softens the grain, hydrates the starch and removes the soluble components that may otherwise contaminate later fractions. The soluble material includes proteins, protein breakdown products, sugars, minerals and B vitamins, all of which will be recovered in a separate process (CIRF, 1959). The protein structure of the endosperm is broken down during steeping, by a natural enzymic process, and this loosens the starch granules and facilitates their subsequent separation.

Steeping leads to a rapid extraction of sugars from the maize and the simpler sugars quickly undergo a natural lactic fermentation. The proliferation of lactic acid bacteria and other less desirable micro-organisms is controlled by the addition of sulphur dioxide, and this also serves to inhibit germination. As the grain moves on through the succession of steeps, the sulphur dioxide level is increased and the lactic acid concentration declines. The sulphur dioxide also has an additional effect on the physical structure of the protein, which swells, becomes globular and is consequently easier to separate.

At the end of the process the steeped corn is drained and then flushed to remove the remaining soluble components, before being dewatered on stationary screens. The steepwater is collected and evaporated from an initial 6-10 to about 50 per cent dry matter. The resulting syrup is primarily a rich source of soluble protein, and the protein content comprises some 45 per cent of the dry matter. Although it contains only a small amount of fermentable sugars, the liquor has a relatively high proportion of lactic acid and more complex sugars. Some syrup is available for use in both the animal feed and industrial fermentation industries, though much of it is blended on-site with maize fibre (see later), and dried, before being marketed as the popular and widely known co-product maize gluten feed. When marketed separately, the syrup is known as either **corn steep liquor (CSL)** or, more correctly, **concentrated corn steep liquor (CCSL)**. The universal use of the term “corn” is in contrast to all the other co-products of maize processing in Britain, which are referred to as maize rather than corn.

The steeped moist grain is then carried in a water stream to a Foos mill, where it is coarsely milled between two studded circular plates, one of which is stationary and the other rotating. The distance between the plates can be adjusted so that the operation imparts a relatively gentle grinding and partial maceration that removes the outer pericarp and breaks the kernels. This allows the subsequent separation of the oil-rich maize germ. The milled corn is flushed through the mills and on to the hydrocyclones where, in the next stage, the whole germs are separated from the other constituents by centrifugal action. The maize germ is washed and then dewatered by screw press before being dried. Subsequently, maize germ is treated in much the same way as any oilseed, being either solvent extracted or screw-pressed to expel the oil. In the larger factories in the United States, maize germ oil is extracted on site but, in Britain, the germ is marketed whole to be extracted elsewhere.

In the UK, the germ produced at both of the British maize fractionation plants is dispatched for oil extraction at the Cargill plant in Hull. This facility, which processes maize germ only from these two sources, is consequently the only source of maize oil and maize germ meal from British processing. At Hull, the germ is first pressed and then solvent extracted to produce an oil that is in demand for the human food market, and **maize germ meal** that is regarded as a good-quality protein source by the animal feed industry. At times, unextracted maize germ has been marketed as a fat-rich supplement for poultry, young pigs and calf diets and, in this form, the oil comprises almost 50 per cent of the feed dry matter. However, the use of the whole germ as a feed - even in such relatively expensive diets – may be expected to be less economic than the extraction of the oil that enables approximately half of the germ to be used in the human diet.

In the maize fractionation plant, after the removal of the germ the remaining material exits as a coarse slurry from the germ hydrocyclones. At this stage, it consists of fibre, gluten, and both free and bound forms of starch, and it is finely ground before being passed through a number of screening processes. These gradually strip the fibre from the other constituents. The fibre is then pressed, mixed with some of the corn steep liquor and flash dried. The resulting mixed product is marketed as **maize gluten feed** – a co-product widely used by ruminant farmers as a source of both protein and energy. Since the protein is largely supplied by the CSL – a rich source of soluble protein - maize gluten feed, despite its name, normally contains little of the insoluble gluten protein.

At times, when the factory driers break down, the two components of maize gluten feed become available separately. CSL is often available in this form but **maize fibre** is less widely known. However, metabolism studies at the ADAS Feed Evaluation Unit (ADAS, 1987) have shown it to be a palatable feed material in its own right. Although the fibre is pressed to around 30 per cent dry matter, it should be noted that this solid fraction has a higher moisture content than the concentrated corn steep liquor.

In the past, a substantial volume of undried maize gluten feed was produced and marketed as animal feed. The attractions to the factory of this moist form of the co-product are immediately obvious: drying costs are avoided. But additional storage and transport are needed for a material that, at roughly half the dry matter content of dried MGF, necessarily occupies greater volume. Marketing flexibility is also reduced, because such a moist material attracts no interest from the compound feed trade and it must be marketed direct to farms. On-farm, the moist co-product is also less flexible than the dried material, and it requires dedicated storage facilities. For these reasons, and undoubtedly other factors – including changes in the relationship between drying and other costs - a decision was made to change factory procedures once more, and return to drying the co-product blend. However, the moist material was widely used and researched, and there are commercial indications that it will return to the market in the Autumn of 2001.

In the processing plant, after the separation of the fibre, the material remaining in the system comprises a mixture of starch and gluten. In contrast to the widely varying size of wheat starch granules, maize starch occurs in relatively uniform granules, and this precludes the need for it to be separated as different starch fractions. Since maize starch particles are heavier than those of the gluten, the two components can be separated by high speed centrifugation. The maize gluten is concentrated in continuous centrifugal separators before being dewatered on a vacuum drum filter and either flash or rotary-dried. Dried gluten is a bright yellow, granular powder that contains some 60-70 per cent crude protein on a dry matter basis. It is marketed as a protein-rich animal feed under the name **maize gluten meal** (not feed).

The maize starch at this point still contains a little gluten and has to be put through a purification process before it becomes available as the finished product. Purified maize starch has a wide range of potential uses – for the food industry, for brewers and for industrial uses in the paper and packaging industries. Much of it is converted into sugars and the resulting syrups are widely used as an alternative to the crystallised product from sugar cane and sugar beet. Some syrups are used in the pet food industry but the product rarely becomes available for livestock feeding – any material found to be outside the required specification is typically reprocessed.

Co-products of maize processing

Corn steep liquor

Corn steep liquor (CSL) is the concentrated form of the extract obtained during the initial stages of the wet milling of maize, and it is marketed as a viscous liquid with approximately 50 per cent dry matter. CSL comprises a mixture of all the water soluble constituents of the grain; it has a rich protein content but is predictably low in both oil and fibre. Its carbohydrate content is somewhat variable because the simple sugars extracted from the maize are converted into lactic acid by natural fermentation. However, the co-product does contain a significant amount of more complex sugars, and this fraction may be boosted by the inclusion of a proportion of the spent wash that remains after the distillation of fermented maize syrup.

The amino acid composition reflects the co-product's cereal origin, and the lysine content, at 1.4 per cent of the dry matter, is relatively low for a material containing 40-45 per cent crude protein in the dry matter. This indicates that CSL may have more potential as a high protein supplement for ruminant rather than non-ruminant stock. The high solubility of the protein implies that it will all be available for microbial synthesis in the rumen though, for optimum benefit, its supply will need to be matched with rapidly available carbohydrate sources. CSL also has a high ash content, amounting to as much as 22 per cent of the dry matter in some samples. This makes the co-product a rich source of phosphorus, particularly for ruminants. But the ash also contains a high level of electrolytes (including a notably high potassium level), and this may make CSL a difficult ingredient for incorporation in pig rations – particularly in liquid diets that may include other sodium-rich feeds.

Corn steep liquor is regarded as a palatable, nutritious liquid for cattle and sheep. It is a common ingredient of molasses-based liquid feeds and can be used as a high protein replacement for molasses in mixed diets for high performance dairy or beef cattle. Some difficulties have been experienced in handling a product which has a tendency to gel and Ewing (1997) notes that, in order to reduce the severity of this problem, CSL is often mixed with molasses before storage.

Maize gluten feed

Maize gluten feed (MGF) is a dried co-product of the maize fractionation industry, and livestock farmers have the choice of purchasing an imported or a home-produced feed material. The British processing industry claims an advantage in terms of the greater palatability of a fresher product - available from Cerestar as a coarse meal, and from Cargill in pelleted form. MGF consists of those parts of the grain that remain after wet milling, and the extraction of the larger part of the starch, the gluten and the germ (CIRF, 1959). Essentially it is a blend of the maize fibre and corn steep liquor – the latter typically provides some 20-30 per cent of the MGF dry matter, but some variation in inclusion rate leads to differences in the chemical composition and nutritive value of the co-product blend. Maize gluten feed from the United States may also contain a proportion of the extracted maize germ but, since none of the germ is extracted on-site at British wet-milling plants, the British co-product does not contain any part of this fraction. Maize gluten feed has been a staple item of the animal feed trade since the 1880's and, with such a long history, MGF may be regarded as a traditional feed for British livestock.

The colour of MGF can vary from fawn to dark brown, and concern has been expressed that the darkest materials may have suffered a degree of overheating, with possibly adverse consequences for protein digestibility. An alternative explanation for the dark colour may be the inclusion of a higher proportion of corn steep liquor, in which a number of Maillard reactions between sugars and amino acids had led to a deeper colour development. ADAS laboratory tests on two samples of MGF, described after visual inspection as "burnt", found pepsin/HCl solubilities of 73.6 and 86.3 per cent, compared with a value of 75.9 per cent for lighter coloured material (ADAS, 1991a). This appears to indicate a range in protein digestibility, but one that is not closely linked to the colour of the feed. A relatively wide range of 66-84 per cent is recorded in MAFF Tables for the *in vivo* protein digestibility of 22 samples of MGF (MAFF, 1990).

Maize gluten feed has a protein level of 20-25 per cent in the dry matter, more than twice the protein content of the original grain. Although this concentration has been largely achieved by the extraction of low protein, energy-rich components such as oil and starch, MGF still contains about 20 per cent starch on average, and this co-product feed is a useful source of both energy and protein for ruminants. However, the starch level can be quite variable, and this may have a direct bearing on the energy value of individual samples (ADAS, 1986a).

The value of MGF for non-ruminant animals is restricted by a fibre content that, on average, is four times as high as that of maize grain. But this component too is highly variable, and a range of NDF values of 33-53 per cent was found in an examination of 10 samples of MGF (ADAS, 1986a). A further factor, which reduces the value of MGF for non-ruminants, is an amino acid composition that is relatively poor in lysine, and its potential is further depressed by low protein digestibility. Apparent ileal digestibility values of 54 per cent for protein and only 48 per cent for lysine are quoted in the Eurolysine Table of feedstuffs for pigs (1988).

Maize gluten feed is suitable for feeding to all classes of ruminant animal, and it may be considered as both an energy and a protein source. Metabolism studies at both the Rowett Research Institute and the ADAS Feed Evaluation Unit have shown a relatively wide range of ME values, and the UK Tables (MAFF, 1990) show a range from 11.3 to 14.2. An average value of 12.9 MJ per kg dry matter suggests an energy source marginally poorer than barley. However, the range of possible ME values suggests that it may be useful to test individual consignments before formulating diets based on large amounts of MGF.

Maize gluten feed has been used at 40 per cent of the diet for intensively-fed beef cross dairy bulls, though no analytical details of the feed were reported (ADAS, 1990). Animal performance was similar to that of other groups of cattle fed on diets containing either rapeseed meal or a proprietary protein concentrate, and a growth rate of 1.45 kg per day was achieved over a 206 day period.

Maize gluten feed has been fed to pigs with satisfactory results. At Terrington (ADAS, 1982), pigs from 35-85 kg liveweight were fed diets containing 0, 10, 20 or 30 per cent MGF that had been formulated to similar DE and lysine levels. No significant differences in growth rate or feed conversion efficiency were recorded. However, there was a progressive decline in backfat thickness as the MGF proportion of the diet increased, indicating the likelihood of either an overestimation of the DE value or a lower efficiency of utilisation of the absorbed nutrients. MGF contains a variable proportion of xanthophylls and the possible effect of these on the development of yellow carcase fat has been suggested (Savery, 1984). However, the xanthophyll level of MGF is only about one-tenth the level in maize gluten meal (Vitec 3a), and no reference to fat colouration was made in the Terrington trial, though the xanthophyll level of the MGF used in this investigation may have been low. It seems reasonable to suggest that the need for a better balanced protein supply in the pig diet may restrict the MGF to an inclusion rate that reduces the risk of this problem occurring.

Moist maize gluten feed

Although not available at the time of writing this chapter, moist maize gluten feed has been marketed in substantial quantities to British livestock farmers, and it could be again. The typical dry matter content of this co-product was 40-45 per cent, and this level gave it a competitive edge over many alternative moist feeds marketed at 20-25 per cent dry matter, in terms of a lower transport cost per tonne of dry matter.

In common with dried MGF, the nutritive value of this moist co-product blend varied in line with the proportions in which the maize fibre and corn steep liquor were mixed, but it proved to be a useful source of both energy and protein for ruminant animals. Alderman (1987) reported a series of digestibility studies in

Holland in which the mean GE and DE values of moist MGF were measured at 18.7 and 15.9 MJ/kg DM, from which he calculated an ME value of 12.9 MJ/kg DM. This ME value is exactly in line with the mean ME value of dried MGF given in UK Tables (MAFF, 1990). Further work was reported in which the organic matter digestibility of wet MGF (70.3 per cent) was shown to be slightly higher than that of dry MGF (67.2 per cent). The difference was shown to be associated with higher cell wall digestibility, and this finding confirmed work by Firkins *et al.* (1985). Earlier work by Firkins *et al.* (1984) had shown both wet and dry MGF to be rapidly degraded in the rumen.

Storage trials have shown that moist MGF stores well, and its composition appears to change little during ensilage (Corporaal and Harmsen, 1984). However, concern has been expressed about its aerobic stability, and Dutch research workers were reported to have routinely added 0.1 per cent propionic acid at ensiling (Alderman, 1987). In the author's experience, such a low level of preservative would be unlikely to prevent aerobic decay (Crawshaw *et al.*, 1980).

Maize gluten meal

Maize gluten meal (MGM) is a dried co-product of the maize fractionation industry, and is a concentrated form of maize gluten that remains after the extraction of starch, germ and bran from the maize grain. MGM has a high protein content, amounting to 60-70 per cent of the dry matter, and this defines the feed as a rich protein source. For non-ruminant species, however, MGM is a poor source of lysine – MGM protein contains only 1.7 per cent lysine, and compares poorly with fishmeal (7.5 per cent) and soyabean meal (6.2 per cent). This relative lack of lysine makes MGM a less than ideal protein source for meat producing animals. MGM is more suitable for laying hens, where methionine is typically the first limiting amino acid. Its methionine content compares well with fishmeal (2.5 v 2.8 per cent respectively) and is substantially higher than the methionine in soyabean protein (1.4 per cent).

MGM has a bright yellow colour, as a consequence of a high xanthophyll content, and, in Britain, its inclusion in laying rations is regarded as desirable because it deepens the colouration of egg yolks. The total xanthophyll content of MGM has been found to vary between 150-300 mg/kg (VITEC 3a) and, where a precise specification of yolk colour is required, it may be necessary to check each consignment of MGM – both for the content of total xanthophylls and for the major components, lutein and zeaxanthin. The same effect in other classes of stock is regarded as less desirable, because it may lead to excessive yellowing of the body fat, and the downgrading of carcases in the slaughterhouse. For this reason the inclusion rate of MGM is usually restricted to 5 per cent in pig and broiler diets.

Since gluten represents the insoluble protein in maize, MGM would not be expected to provide a rapidly available protein for ruminant animals. In fact, it appears to be a good source of undegradable protein (Chalupa *et al.*, 1999) – resistant to degradation in the rumen and a good source of methionine from the small intestine. This could represent a niche role for MGM in dairy cow rations, because methionine has been identified as one of the two most limiting amino acids for the synthesis of milk protein (Schwab and Satter, 1976). However, since the other limiting amino acid is lysine, MGM would appear to provide only a partial solution.

Broken maize / maize screenings

All of the maize grain used in the UK is imported and is subjected to considerable handling before it reaches the processing factories. A proportion of broken grain is thus inevitable and, because this is potentially much less valuable to the processor, it is screened out immediately and consigned to the animal feed market. The screenings also include a proportion of small, immature grain. The composition of these maize screenings may be expected to be very similar to that of the whole grain, but there are differences. Maize fragments tend to contain much of the fibre and the starchy grits, but little of the germ and these differences result in a feed material that has more fibre and less oil than the whole grain. However, the feeding value of this cereal co-product is largely determined by its starch content and, at more than 70 per cent of the dry matter, broken maize is clearly a high-energy feed. This feed could be used for any class of stock, but it is particularly suitable for high density formulations and currently, much of it is used in poultry diets.

Maize fibre

Maize fibre is a co-product of maize fractionation that is usually available only in blended and dried form as a component of maize gluten feed (MGF). The co-product is separated from the gluten and starch as part of the wet milling procedure, and is then pressed to a dry matter content of 35-40 per cent. Maize fibre is usually re-combined with corn steep liquor before being dried, although this blend has also been available as a moist feed. Maize fibre has been marketed as a separate co-product, but at present it is available separately only when there is a breakdown at the factory. Maize fibre has a bright yellow appearance, and its consistency has been described as similar to that of damp wood shavings.

As its name implies, maize fibre is a high-fibre material that is more suitable as a feed for ruminant rather than non-ruminant animals. However, a starch content that averages about 18 per cent of the dry matter means that this feed is more than just a source of fibre. Five samples of maize fibre were assessed at the ADAS Feed Evaluation Unit (ADAS, 1987), and all parts of the feed were found to have a digestibility of more than 70 per cent. Fibre digestibility was notably high, and NDF digestibility averaged 77 per cent. This indicates that the material could have a roughage-sparing role in intensive diets, such as those used for high yielding dairy cows, though an additional supply of long forage would be needed. The ME value, determined *in vivo* at the FEU, ranged from 12.2 to 14.5 MJ/kg DM. Such variability is in line with that found in a larger study of maize gluten feed (MAFF, 1990) though, in contrast to the MGF, there was no apparent link to differences in composition. The mean ME value of 13.4 appears high, though this is largely a function of the relatively high gross energy value of 20.3 MJ/kg DM that was measured in these five consignments.

The crude protein content also appears variable; MAFF Tables quote values ranging from 11 to 21 per cent of the dry matter. Since the soluble protein was removed in the initial steeping, and the remaining protein is associated with the fibrous part of the plant, a mean digestibility of 72 per cent would seem moderately good. The ash content of maize fibre is low, and soluble minerals such as sodium and potassium are predictably low.

Maize fibre has proved to be a palatable feed for ruminant stock and is probably best incorporated into a mixed diet. The feed may be expected to provide digestible fibre, slowly degradable starch and some by-pass protein - though its relatively poor lysine and methionine content would not make it an ideal source of DUP.

Maize germ meal

The germ is one of the most distinctive features of maize grain. Though it contributes only about 6.5-7 per cent of the fresh weight of the kernel, its high oil content provides significant enrichment of the oil content of the whole grain. Unlike other fractions of the maize grain, which are closely bound and separated only with difficulty during wet milling, the germ is a distinct entity that can be separated more easily. At the two UK processing plants, the separated germ is washed, pressed and dried before dispatch to the Cargill oil extraction facility at Hull. This plant is the sole source of maize oil and maize germ meal from British processors.

The oil, which constitutes about 50 per cent of the dried germ, is highly valued by the human food industry, and thus its effective extraction is desirable. Old processing methods that relied on pressing or even double-pressing of the germ tended to leave a significant proportion of oil in the residual cake. At Hull, these methods have now been replaced by single pressing followed by solvent extraction, and the oil content of the resulting meal has been reduced to no more than 3 per cent. Oil extraction effectively doubles the concentration of the other constituents of the germ, and this lifts the protein content to 22-23 per cent and the crude fibre level to 9-10 per cent, though the ash content remains relatively low. That leaves approximately 50 per cent of the meal to be made up of carbohydrates, of which starch is almost half.

The composition given in the preceding paragraph relates to the maize germ meal currently being produced by the Hull factory (S Arundel, Cargill; personal communication). It is quite different from data quoted in the UK Tables (1990), which include a range of oil (ether extract) content of 5.4-12.9 per cent of the dry matter. The Roche VITEC 3 compendium of Animal Nutrition and Vitamin News, published between 1988 and 1990 (VITEC 3b) also refers to "great variation between different sources of maize germ meal", with oil contents quoted of 1.0 to 21.9 per cent of the fresh weight. If this position still holds it must relate to imported material, since both the separation of the germ and its extraction in British factories appear to produce a relatively consistent co-product.

Maize germ meal is pale yellow in colour and is a poor source of xanthophyll. This implies that it has little value as a supplier of pigment for egg yolk colouration, but may be expected to have no undesirable consequences for body fat colour in meat animals. VITEC 3 refers to the impact of unsaturated oil in maize germ meal on the softness of carcase fat, and asserts that this makes the co-product undesirable in pig feeds or feeds for other meat animals. However, the much more efficient extraction of oil that is achieved by modern processing, must render this comment redundant – at least in relation to the British co-product. In any case, maize oil does not appear to be a richer source of unsaturated fatty acids than sunflower or soya oils.

As a protein source, maize germ meal has a similar protein content to maize gluten feed, but its lysine content is superior. Although it has a much lower protein content than maize gluten meal, maize germ protein is much richer in lysine.

8 Milk Processing

Milk is widely regarded as being close to the ideal food; it is rich in protein, fat, carbohydrate and major minerals, and a valuable source of both fat and water soluble vitamins. Thus the National Dairy Council's description of milk (NDC, 2000) as (merely) "one of our most nutritionally complete foods, but lacking fibre and starch, being a poor source of iron and vitamin D, and containing relatively little vitamin C", seems somewhat niggardly. Animal feed specialists would consider milk as merely a part of the diet for all but the very youngest animals, and they consequently value the material for its many positive features.

More than 14 million tonnes of milk are produced in the UK each year, and each consignment is checked and sampled on-farm before collection. Such individual testing ensures total traceability of supply, and any quality problems can be rapidly traced to source and the affected milk excluded. Additional weekly tests, taken at random each week, ensure that the milk is clean and free from contamination. Milk removed from healthy udders contains only a low level of bacteria, and careful attention to the outside of the teats and to the milking equipment, delivery lines and bulk tanks in the parlour and dairy, should ensure a low bacterial count in milk leaving the farm. A maximum limit of 100,000 bacteria per ml of milk is imposed by legislation but, in practice, typical bacterial counts in UK milk are less than 20,000 cells per ml (NDC, 2000).

Within every dairy herd, the cows are subject to a number of mandatory checks to assure freedom from tuberculosis, brucellosis, and enzootic bovine leucosis. Mastitis controls imposed over the last few years have dramatically curtailed infections but, when it does occur, the milk from any cows treated with antibiotics is discarded for several days to avoid any traces appearing in milk delivered to the manufacturing plant. Bovine spongiform encephalopathy measures are also in place, and are expected to lead to the eradication of this disease within the next few years. There has never been any evidence of the transmission of this disease by way of the milk, even in tests with calves consuming milk from BSE-affected cows (FSA, 2000).

Bulk milk collection tankers are cleaned according to a strict Code of Practice and this is supplemented by regular monitoring of tanker hygiene. Together with improvements in quality on the farm, this emphasis on hygiene has enabled daily collections to be replaced by alternate day collection, without impairing the quality of the milk supply. Once the milk reaches the dairy it is checked again for appearance, smell and hygienic quality to ensure continued freedom from contamination. Over 99 per cent of the fresh milk supply, and all of the milk that will be processed into the wide range of dairy products, are heat treated at the start of the process (NDC, 2000). The vast bulk of the milk received by the milk processor will be pasteurised, by maintaining at a temperature of at least 71.7°C for a minimum period of 15 seconds. This pre-treatment kills any pathogenic micro-organisms, and reduces the number of spoilage organisms that may be present, thereby extending the milk's shelf-life.

Approximately half the UK milk supply is marketed in liquid form to the human food market. The other half is converted into a wide range of dairy products, the major features of which are shown in Table 1.

Table 1: Major Milk Products

Product type	*% of Manufactured Milk*
Cheese	48
Milk powder	28
Condensed Milk	10
Butter	4
Cream	4
Others	6

Source: NDC (1999)

Collectively these fractions account for one hundred per cent of the manufactured milk, and the completeness of these figures may appear to imply that there are no co-products of milk processing. However, many of the products in Table 1 represent only a selected portion of the milk that is processed for the purpose, and there is an associated "other portion". Traditionally, the other portion was used for pig feeding - often on the farm where processing was taking place – and both skimmed milk from butter churning and whey from farmhouse cheese manufacture were widely known feeds of good repute. Over the last thirty years the dairy industry has increasingly recovered much of this "other portion", converting it into higher value products with a consequent reduction in the amount that has been available at modest prices as liquid animal feeds.

Milk co-products have traditionally been fed to pigs
(courtesy: Roquette UK)

The key development that enabled the dairy industry to refine and reduce its co-product output was the introduction of membrane filtration (Nielsen, 1992). This technology, spanning a range of processes from microfiltration, through ultrafiltration and nanofiltration to reverse osmosis, enables molecules and ions of various sizes to be separated. Thus fat globules can be separated from casein micelles by microfiltration, and ultrafiltration can separate whey proteins from the lactose and mineral fractions. With finer membranes and the application of greater pressure, nanofiltration may then be used to remove sodium, potassium and chloride ions and thereby concentrate the remaining lactose. In the finest procedure, reverse osmosis may be used to remove water and to concentrate the solids in a number of dairy products including milk, skimmed milk and whey.

All of these filtration processes yield two fractions: the concentrate (or retentate) and the permeate (or filtrate), and the ultimate objective of the dairy industry is that both fractions in their isolated forms may be marketed as dairy products. Although these markets would include specific animal feed uses, such as the production of milk substitutes for calves and occasionally for other infant stock, this aspiration, if fully achieved, would deny the feed industry the liquid co-products that it has traditionally used. That position has yet to be reached, though it is difficult to determine the size of the residual milk co-product market. Somewhat oddly, the author has found the dairy industry reluctant to quantify or to discuss liquid co-products, as if this residual market were an embarrassment. Yet the scrupulous attention to sourcing and to hygiene, that is the hallmark of all aspects of milk production, collection and processing, gives these co-product feeds an enviable provenance that should satisfy the most rigorous feed inspection. The co-product information presented in this chapter has largely been obtained from the hauliers who supply a range of dairy co-products to pig farms and, collectively, their efforts account for approximately 500,000 tonnes per annum of liquid dairy co-product.

Cheese manufacture

Cheese accounts for the largest fraction of manufacturing milk, and it seems inevitable that the prime source of milk co-products for animal feeding will be whey. MAFF Statistics on the production and use of whey (Alison Bromley, personal communication) show that, throughout the last decade, almost three million tonnes of whey have been produced by UK dairies each year, but less than 25 per cent has been accounted for by the production of whey powder. Some of the remaining two million tonnes is processed in a number of ways; by evaporation, centrigugation or reverse osmosis to produce **whey concentrate**, by the partial removal of the sugar fraction to leave **delactosed whey** or by the removal of protein to yield **whey permeate**. A proportion of all of these liquid co-products is made available to pig farmers, in addition to a supply of **whey** in its unprocessed form, although the volume of this latter material is declining. Some of the recipient farms are owned and managed by the cheese factory, and it is consequently difficult to determine the amount of dairy co-products used in these situations. It may even be possible to argue that, in such vertically integrated operations, the co-product of milk processing is not another fraction of milk but pigmeat, though that may be equally difficult to quantify. However, not all cheese factories run their own pig units and, where they do, unless the whey is dried on-site, there is usually a surplus of a whey co-product to be marketed as animal feed.

In cheese manufacture, differences in processing conditions give rise to a great range of cheese types, but the basic principles of cheese making remain the same – as they have done for over a thousand years (NDC, 2000). Differences mainly occur after the initial separation of the whey and, viewed from the co-product angle, the greatest variation between cheese types is likely to be in the proportion of whey that is separated.

In virtually all factory operations, pasteurised milk is rapidly cooled to about 31°C and transferred to the cheese vat. Selected laboratory-grown cultures of lactococcus lactis are added to convert the milk sugar into lactic acid. After a short period, rennet is added to the acidifying milk and the enzyme it contains (chymosin) brings about a coagulation of the casein fraction, and an entrapment of the milk fat in the curd. Traditionally, the rennet was obtained from the stomach of slaughtered calves, but vegetarian alternatives are now available – either in the form of microbial coagulants or as a pure enzyme produced by a modified organism. Vegetable colourants may also be added at this stage. After 40 minutes or so, the curd is firm enough to be cut and this action helps to release the liquid whey. The curds and whey are then "scalded" – a process of stirring, with a slow increase in temperature to a maximum of about 39°C, which effects a change in the texture of the curd whilst acidity continues to develop. Scalding encourages the further release of whey from the solid curd, and after settling – or pitching – the whey is run off from the curd. In hard cheese manufacture, the solid material is further processed by cutting and stacking, and the resulting pressure leads to the release of more whey and greater acidification – at the end of this process, the whey will have developed a pH level of approximately 5.2. In the production of soft cheeses, there is less pressing and a lower whey output. Cottage cheese manufacture includes a double washing of the curd, which may be expected to dilute the whey. However, the additional water removes some of the curd and thus the net effect is likely to be simply a greater ouput of whey.

Whey continues to be released as the curd is first milled and then pressed into the blocks, in which form hard cheeses will be matured and eventually marketed. However, there is an additional step that has a significant effect on the composition of the whey, and that is salting. Salt is added at approximately 3 per cent to the milled curd prior to pressing, to assist preservation and bring out the flavour of the cheese. Salt is also incorporated in most other cheeses, although their salt content does vary, and so does that of the associated whey. Cheese manufacture involves the removal of more than half the organic matter of the original milk, and the mineral content of the remaining solids is thereby increased. However, it is the additional salting of the cheese that is responsible for much of the sodium that appears in whey. Since sodium chloride is highly soluble, it is inevitable that much of the added salt will appear in the liquid fraction that is separated subsequent to this stage. Since a variable amount of whey is removed after salting, and added to that separated earlier, this process leads to variability in the sodium content of the whey from different sources, and this may be of nutritional significance. Nutritionists operating in the co-product pig feed market routinely monitor the sodium content of feeds.

The entire chain of production from milk to matured cheese is controlled by the Dairy Products Hygiene Regulations 1995. However, quality is most effectively ensured by the factories themselves, and by their adoption of QA systems aimed at achieving good manufacturing practice. The high standards that they set, which are appropriate to the production of high quality food products, are also applied to the associated co-products. They provide a commendable provenance for the re-entry of dairy co-products into the food chain via animal feeds.

Other milk products

Whilst some milk manufacturing processes include little scope for co-products, because either the primary or the secondary product is dried into a type of milk powder, the "Others" section of Table 1 offers some possibilities.

Yoghurt production in the UK has increased five-fold over the last 30 years (Alison Bromley, MAFF Statistics; personal communication) and now utilises more than 250,000 tonnes of milk per year. The total volume of yoghurt produced from UK milk may exceed this figure, by virtue of the addition of fruit, nuts and sugar, although the milk is initially concentrated to around 16 per cent solids, by evaporation or the addition of milk solids. This concentration improves the texture of the yoghurt and reduces the risk of the product separating into layers. Some of this flavoured yoghurt becomes available to the animal feed market, but the dry matter content of the co-product material is typically much lower. This reflects the inclusion of water which has been used to wash the production line between different yoghurt types. It is the rich variety of yoghurt flavours, and the consequent need to flush the system continually to preclude cross contamination from one to another, that gives rise to the availability of co-product yoghurt for animal feeding.

Ice cream has been known and enjoyed for more than 400 years (McGee, 1984). It comprises a mixture of milk solids, sugar and other minor components and may contain vegetable oil in place of milk fat. The texture is achieved by freezing and aerating a liquid blend of ingredients, but this is of little relevance to the animal feed market since the volume of co-product delivered to farm is in liquid form. Major variations in ice cream quality are to be found in the proportions of air and fat – luxury versions tend to have twice the fat but half the air content of cheaper versions (McGee, 1984). Ice cream for animal feeding will inevitably comprise a blend of the different types and will exhibit less variation – particularly when it is consistently obtained from the same source. Of significance to human and animal markets alike is the legal requirement (Dairy Product Hygiene Regulations 1995) that the ice cream blend must be heat treated to kill any harmful micro-organisms and then rapidly cooled prior to freezing. Thus the microbiological quality of both product and co-product is ensured throughout the chain that leads from milking machine cluster to ice-cream scoop.

Co-products of milk processing

Whey

Whey is a liquid co-product of cheese manufacture that has long been used and valued as a high quality pig feed; Schingoethe (1976) refers to whey being fed to the swine in ancient Rome. Today, much of the whey produced in the larger cheese factories is further processed, and very little is made available in liquid form for animal feeding. However, despite this trend, whey in its various forms remains the predominant liquid feed that is widely available to the UK pig industry, though its stronghold is in the South-West where milk production and relatively small-scale cheese manufacturing activities are concentrated.

The principal feature of whey is its high water content, typically comprising 93-95 per cent of the co-product. This inevitably results in a high haulage charge on the dry matter fraction and a subsidy is needed to render the feed competitive. The inclusion of a substantial amount of whey in the pigs' diet may also lead to the production of a mixed feed of low dry matter content and, in consequence, the possibility of reduced feed intake and animal performance (Barber, 1998). To preclude this eventuality, an alternative solution may be to limit the use of otherwise competitively-priced liquid and moist feed alternatives. In this situation, the pig farmer would have the choice of a least-cost diet of low dry matter, or a more expensive diet that satisfies his requirements in terms of minimum dry matter content.

The high sodium content of whey, typically 1-1.3 per cent of the dry matter though occasionally higher, may give rise to further problems. Unless balanced with low sodium feed ingredients – and liquid feeders often have a number of other sodium-rich feeds available - this may lead to low feed intake and reduced performance, as well as a problem with dirty pigs. Salt toxicity is primarily regarded as a problem of water deprivation, but it is important to appreciate the impact that additional dietary sodium has on water consumption and urine output. ARC (1981) suggests that for every 10g of sodium chloride (3.9g sodium) above the requirement level, the pig needs to consume and to excrete an additional litre of water.

Other than these considerations of water balance, whey may be regarded as a highly digestible feed source. Lactose, at more than 70 per cent, is the principal constituent of the dry matter although, on-farm, much of the sugar may be fermented to lactic acid (and a smaller proportion of volatile fatty acids), and the pH of the whey may drop to 3.25-4.0. Mitchell and Sedgewick (1963) noted that the development of acidity occurred rapidly during summer, but more slowly at the temperatures that typically prevail during a British winter. Table 2 reports the results of a study carried out in the author's laboratory (Crawshaw *et al.*, 1976) of the changes that occurred in whey held for seven days at different temperatures:-

Table 2: Changes occurring in whey stored at different temperatures

	pH			*Lactose %*			*Lactic acid %*		
Temp.	*1°C*	*20°C*	*Ambient*	*1°C*	*20°C*	*Ambient*	*1°C*	*20°C*	*Ambient*
Day 0	5.4	5.4	5.4	4.71	4.71	4.71	0.22	0.22	0.22
Day 1	5.4	4.7	5.2	4.71	4.37	4.78	0.21	0.48	0.31
Day 2	5.4	4.6	5.2	4.78	4.31	4.78	0.21	0.58	0.34
Day 3	5.4	4.3	5.1	4.71	4.17	4.58	0.21	0.62	0.32
Day 4	5.4	4.0	5.1	4.71	3.77	4.44	0.21	0.68	0.38
Day 7	5.4	3.9	4.9	4.71	2.69	3.70	0.19	0.68	0.41

The 1°C samples were held in the refrigerator, the 20°C samples in a constant temperature room and the ambient samples were stored outdoors where the temperature during the 7 days varied between 1 and 20°C.

At low temperature, the whey remained unchanged throughout the storage period, but at a constant 20°C almost half of the lactose disappeared within seven days and the lactic acid concentration increased threefold. It is notable that these measurements were carried out on fresh whey that had been stored in clean containers. Acidity may be expected to develop more rapidly where a new consignment is added to the residue of the previous load, as would often occur in commercial practice, since this would result in the inoculation of the fresh whey with an actively multiplying microflora.

As may be seen from Table 2, the loss of lactose exceeded the increase in lactic acid content by a considerable margin. Although no attempt was made in this study to assess the energetic consequences of these changes, the possibility of a significant loss of digestible energy cannot be ruled out. Fox and O'Connor (1969) noted the involvement of other micro-organisms in the souring of whey, and particularly the proliferation of yeasts, which would lead to the development of alcohol and the loss of carbon dioxide. However, despite such changes, experience has shown that soured whey remains acceptable to pigs, and its nutritional value is similar to that of sweet (formalin-preserved) whey, at inclusion rates of up to 30 per cent of the diet, though not at higher feeding levels (Barber *et al.*, 1978).

There may even be advantages to feeding whey in this acidified form. Whey bloat, diarrhoea and death have occurred in situations where lactose consumption has exceeded the pig's ability to digest the sugar in the small intestine. This possibility increases as the pig gets older and its lactose tolerance falls in line with a decline in the concentration of intestinal lactase (Todd, 1977; Beames and Taylor, 1991). The risk is avoided where lactic acid replaces lactose. The presence of lactic acid in the intestine may also have additional prophylactic value, by promoting conditions that favour beneficial micro-organisms and discouraging the growth of undesirable organisms. Cole and colleagues (1968) reported a reduction in *E Coli* numbers in the duodenum and jejunum when lactic acid was added to the diet. Brooks (1999), quoting information from a survey in the Netherlands, noted that the incidence of sub-clinical salmonella infection was ten times lower on farms with liquid feeding, and was particularly low on farms that fed acidified whey.

Despite the loss of the casein that was removed during cheese making, the quality of the remaining whey proteins is good. Although the crude protein level is only 15 per cent of the dry matter, the lysine content is 1 per cent and Schingoethe (1976) claimed that pig farmers traditionally used whey as the sole protein supplement to barley for fattening pigs. Such a diet would seem unlikely to support optimum growth rates in modern genotypes and, as long ago as 1969, Hanrahan reported leaner carcases associated with the use of a compound feed supplement to whey that contained 16 as opposed to 14 per cent protein. Both of these protein levels are significantly higher and the quality better than typical barley protein.

Whey is also notable as a good source of calcium and phosphorus, the two mineral elements required in greatest amount by the pig at all stages of production. Furthermore, whereas the phosphorus content of cereals and oilseed meals is poorly available, the phosphorus availability of dairy co-products such as whey is relatively high (NRC, 1998). Jongbloed and colleagues (1991) quoted phosphorus digestibility values obtained in the Netherlands for a number of feeds, ranging from 10 per cent for tapioca meal and 17 per cent for maize up to 82 per cent for whey powder and 91 per cent for skimmed milk. Such a high bio-availability for the phosphorus in dairy co-products is useful not only to the pig but to the environment too, since it has been established that the phosphorus content of pig effluents leads to the undesirable eutrophication of ground water and fresh water sources (Gerritse and Zugec, 1977).

Whey can be fed to pigs of almost any age and dried whey is often included in creep feeds offered to pigs before weaning. Beames and Taylor (1991) stated that 20 per cent whey powder was a standard recommendation for weaner diets, but suggested that a higher proportion may be used in liquid diets. As noted previously, Barber *et al.* (1978) recorded satisfactory results with up to 30 per cent dietary whey. However, the optimum inclusion rate under practical conditions may be lower, if other competitvely-priced liquid feeds are also to be incorporated in a diet without burdening the pig with an undesirably high intake of either sodium or water.

Whey co-products are easily fed through a pipeline feeding system
(courtesy: Meyer-Lohne (UK))

Whey may also be used for cattle feeding. In a review of United States experience, Schingoethe (1976) concluded that ruminants – both beef and dairy cattle - may consume up to 30 per cent of their dry matter intake as liquid whey without impairing performance. Milk production was not affected when liquid whey replaced all or part of the water offered to lactating cows (Anderson *et al.*, 1974). This finding accords with the author's own experience, though it was found necessary to switch off the water supply in order to induce the cows to consume a diluted supply of fresh whey (Griffiths and Crawshaw, 1977). After building up the whey to full strength over a period of 10-14 days, the water supply was switched on again and the cows were then observed to drink from both sources. It should be noted that, in the author's field observation study, despite the use of a daily supply of fresh whey, the average consumption amounted to only 15 per cent of the total dietary dry matter – half the amount suggested in the studies reported by Schingoethe. It seems likely that the voluntary consumption of whey will be affected by the other constituents of the diet; at least some of the American studies used dry rations based on hay and grain, whereas the UK study included ad libitum good quality grass silage.

A few problems have been encountered when feeding whey to cattle. In particular, whey feeding causes an increased volume of urine, and some scouring which leads to a greater proportion of dirty cows, with potentially negative implications for milk hygiene and the occurrence of mastitis. One report (Welch *et al.*, 1973) also mentioned teeth erosion, but this has not been confirmed in other studies of eight to ten months duration with heifers and steers (Schingoethe, 1976).

Whey concentrate

Whey concentrate is a high quality dairy co-product produced from whey by the process of reverse osmosis or more traditionally by evaporation. This final processing stage takes a low-value liquid of perhaps 5-7 per cent dry matter and transforms it by the removal of water into a nutrient-rich fluid, with a dry matter content of 30, 40 or even 50 per cent. In this concentrated form, the co-product may be further processed by the milk factory, hauled to another processing site, or made available for use as a liquid feed by pigs.

Whey concentrate has all the virtues of unmodified whey; it is a highly digestible, energy-rich feed with a medium protein content of good nutritional quality. But in contrast to the unmodified co-product, these positive features are not burdened by the cost of transporting a much greater volume of water. Whey concentrate is a cost- effective feed that should (at the prevailing prices in 2001) require no subsidy to be competitive on the animal feed market. The higher dry matter content is also attractive to the farmer, since the co-product may be fed without concern for its bulkiness. A feed containing 30 per cent dry matter may be regarded as the equivalent of 1 part meal to 2.2 parts water, which is well within practical guidelines for a liquid diet (Barber, 1998). In fact, when offered a feed/water ratio of this order, the pig may be expected to drink additional water and adjust its own water balance. This is clearly advantageous, and it also affords the opportunity for the pig to compensate for a variable sodium content in the diet.

Concentrated whey is more stable than the dilute form of the co-product, because the decreased water availability (increased osmotic pressure) severely restricts microbial growth. Hanrahan (1977) studied the compositional changes that occurred during the storage of fresh whey and two concentrated forms containing either 20 or 30 per cent total solids. Within a week the lactose had disappeared from the fresh whey, but the two concentrated co-products maintained their lactose concentration for 8 and 17 weeks respectively. Thus the storage of concentrated whey may be regarded as energetically more efficient, but the possibility that this high lactose feed may lead to a greater risk of whey bloat (than soured whey) must be borne in mind when formulating rations for finishing pigs.

Whey permeate

Whey permeate is the deproteinised form of whey that is becoming available in increasing amount at the present time. Owing to the great demand for whey protein, for use in baby pig and other specialist feeds, milk processors are reworking more of the whey to recover the protein fraction. Whey permeate is essentially the residual mixture of lactose, minerals and water, though it does contain a small proportion of (largely) non-protein nitrogen. In its natural state, whey permeate may be expected to have a very low dry matter content, of less than 5 per cent. Some of the smaller factories have a return arrangement with their milk suppliers, whereby this dilute co-product is returned to the farm – possibly for use as a liquid feed for ruminant animals. Larger factories typically concentrate the permeate, usually by evaporation, before offering it as a higher value material for the animal feed market. From each cheese factory, the whey permeate typically has a consistent dry matter content, but the dry matter level does vary between factories. There are sources where the co-product has a dry matter content as low as 16 or 18 per cent, and others where it is available at 22 and 25 per cent. A **concentrated whey permeate** is also produced in some factories with a dry matter content of 45 per cent. Although offering further advantages in terms of reduced transport cost – per tonne of dry matter - this latter material can be difficult to handle because of a tendency for some of the lactose to crystallise out of solution.

Lactose is the predominant constituent of the dry matter, comprising some 83 per cent. The residual protein content is low, and is typically less than 4 per cent of the dry matter. The mineral fraction is further concentrated by the removal of the protein, but only marginally. After adjustment of the diet to balance the reduced protein content, this co-product can be used as a direct replacement for whey. The higher dry matter forms of whey permeate will enable a higher dry matter diet to be formulated, and free access to water should enable the pig to cope satisfactorily with the sodium content of this feed. At 16-18, and more so at 25 per cent dry matter, whey permeate may be expected to be more stable than the much more dilute fresh co-product, and the lactic acid concentration will develop more slowly.

Delactosed whey

The name "delactosed whey" may appear to imply that this is a dairy co-product from which almost all of the lactose has been removed, but this is certainly not the case; in fact lactose remains the principal constituent of the dry matter. The feed could more accurately be termed lactose-reduced whey because, compared with unprocessed whey, the lactose content has been reduced from 75 to approximately 55 per cent of the dry matter. It is technically possible to reduce the lactose content of whey more substantially but, compared to whey protein, the lactose has a much lower monetary valuc. This reality makes it more difficulty to justify investment in the necessary processing equipment. At different factories, delactosed whey has as much as 60 or as little as 50 per cent lactose.

The consequence of the limited separation of lactose is an increase in the other fractions, namely the protein and the minerals. Thus the protein content of delactosed whey is 20-24 per cent in the dry matter, and such a level of high quality protein makes the feed a potentially valuable supplement for growing pigs. The sodium content at around 2.4 per cent is a less desirable feature and restricts the amount of delactosed whey that can be recommended in the pig's diet. The water content of the feed should present no problem, since delactosed whey is typically concentrated to 38-43 per cent dry matter.

Ice cream

Ice cream is a liquid, dairy co-product that is available from ice cream factories. Feed supplies arise at the start and end of each production run, and as a consequence of product changes when different flavours and specifications become mixed. Feed ice cream may also include a proportion of the water that has been used to flush the production line between the manufacture of different flavours.

Ice cream is essentially a blend of milk fat or vegetable oil with sugar and milk solids, though it may also contain a significant proportion of fruit and a number of other minor ingredients. The dry matter content of the co-product is variable, though considerably lower than the 38 per cent found in dairy ice cream on the human market (NDC, 2000). However, the dry matter is highly digestible and, with a fat inclusion of perhaps 20 per cent and a sugar content of more than 50 per cent, ice cream is a feed of very high digestible energy value. The sugar content makes the feed prone to rapid fermentation, with the production of both lactic acid and carbon dioxide. This inevitably leads to a loss of feeding value during storage, though the feed remains highly palatable. The protein level is relatively low, at around 9-10 per cent of the dry matter, but since the protein derives from milk solids, it may be expected to have a good amino acid balance.

Ice cream is used as a high-energy ingredient in liquid diets for pigs. Its high oil content and high energy value must be carefully balanced, and the dietary inclusion rate would not usually exceed 10 per cent.

Yoghurt

Yoghurt is a fermented dairy co-product which, like ice cream, becomes available as a liquid animal feed at the start and end of production runs in the factory, and during the changeover between different product types. A wide range of flavours is available in both low fat and "rich and creamy" types, with the fat content varying more than fourfold from approximately 0.7-0.8 to 3.4 per cent in fresh yoghurt. The dry matter content is also variable, from 15 per cent in plain yoghurt up to 23 per cent in fruit yoghurts. These twin variables have a major impact on the energy value of the product that is sold in the human market, and a wide range of energy values can be found, from 56 to 530 kJ per 100g.

This variability is reduced in loads that are consigned to animal feed, because they tend to comprise a mix of different product types. But it is useful to bear in mind the potential range in energy value and to establish the typical range for any source. Of great significance to the feed market is the dry matter content, which is typically much lower than that of the food product. The inclusion of flushing waters in co-product yoghurt reduces the dry matter content to between 5 and 9 per cent.

For the liquid-feed pig market, in which feed yoghurt is principally used, it is primarily a protein source. Since yoghurt is developed from milk, into which additional casein and other milk solids may have been included, the protein quality is also good. The protein content of plain yoghurt is diluted by the additon of fruit and sugar, but the co-product may be expected to have a protein content within the range 18-34 per cent on a dry matter basis (NDC, 2000). Additional benefits may also be claimed, such as the probiotic effect associated with the presence of lactic acid bacteria in yoghurt. As noted in the section on whey feeding, a reduction in

E Coli numbers and the incidence of sub-clinical salmonella infection has been associated with acidified whey (Cole *et al.,* 1968; Brooks, 1999). Before the co-product leaves the factory, the conversion of some of the lactose to lactic acid will help to safeguard the pigs from whey bloat, a potentially lethal condition which may develop as a result of the fermentation of lactose in the hind gut.

Salvage milk

Milk is also available from milk packaging plants, and represents a tiny fraction of the milk that is used to supply the liquid milk market. A typical analysis confirms that the product is skimmed milk, with a dry matter content of 9 per cent and a fat content of only 0.1 per cent. As such, it represents a high value feed that was once widely available and highly valued. Typical salvage milk has a protein content equivalent to 37 per cent of its dry matter, and this is high enough for the co-product to be used as the sole source of protein in a cereal-based diet. The mineral fraction contains highly available forms of calcium and phosphorus, and the high bio-availability of phosphorus is of particular note. Comparison with vegetable phosphorus sources has shown the phosphorus content of skimmed milk to be up to nine times more available (Jongbloed *et al.*, 1991).

Since lactose constitutes more than 50 per cent of the dry matter, there may be some risk of "whey bloat" if large amounts of milk are consumed – particularly by the older pig. However, lactic acid souring may be expected to develop during storage, and this should reduce the risk.

9 Potato Processing

Potatoes have been part of the human diet for at least two thousand years. They are believed to have originated in South America, where they were eaten by the inhabitants of the area now known as Peru. Potatoes were unknown in Europe until the sixteenth century, when they were brought back from the New World by adventurers such as Sir Walter Raleigh and Sir Francis Drake, although it is believed that potatoes arrived in Spain before they reached Britain. They have been used as a staple food in the British diet for the last two hundred years.

The supply of potatoes for animal feeding followed their use in the human diet. Any surplus stocks for which there was little demand, and tubers regarded as too small, too large or too irregularly shaped to be marketed commercially, have traditionally been fed to farm animals - ruminants, pigs and even chickens. However, it has long been recognised that, for non-ruminant animals, raw potato has a significantly lower feeding value than cooked potato (Kellner, 1908). Partly this difference can be explained by a slight improvement in starch digestibility, which follows gelatinisation during cooking, but the major difference is in protein utilisation (Whittemore *et al.*, 1973). Raw potato has been found to contain an anti-nutritional factor that interferes with protein digestion – not only of the potato protein but of the protein in the whole diet – but this factor is inactivated by cooking (Whittemore *et al.*, 1975). Raw potatoes do not present such a problem to cattle and sheep since the anti-nutritional factor is inactivated by microbial action in the rumen. However, there are other potential problems; caution is still needed when feeding whole potatoes because of the risk of choking, and chopping or slicing the potatoes before feeding is often recommended in order to reduce this danger. Thus the co-products of potato processing may be regarded as innately safer than the original tubers. A further feeding problem can occur as a result of soil contamination when potatoes are harvested in wet conditions – animals offered dirty potatoes typically consume smaller amounts and animal performance is consequently reduced (ADAS, 1986b).

During the second half of the twentieth century, an increasing proportion of the potatoes purchased for human consumption was processed by the food industry into a range of convenience foods. By 1999, this proportion had grown to a substantial 43 per cent and was still increasing. However, not all of the potato products consumed in Britain are processed here, and it is only this latter fraction that leads to the availability of co-product feedingstuffs for British livestock. Early figures showing the development of potato processing in Britain over the last fifty years appear to be unavailable, but Table 1 demonstrates the industry's continued growth during the 1990's.

Table 1: Potatoes for human consumption in Great Britain

	1993/94		*1998/99*	
	'000 tonnes	*%*	*'000 tonnes*	*%*
Frozen or Chilled	709	47.5	1177	59.0
Crisped	601	40.2	732	36.7
Canned, Dehydrated and Others	184	12.3	85	4.3
Total Processed	1494	100.0	1994	100.0

Source: British Potato Council

Potato processing is dominated by two major industries - frozen food production and crisp manufacture - with canning, dehydration and a range of other prepared food products making a smaller and declining contribution. Popular terminology may be a little confusing to an international readership, with the principal products of these major industries known in Britain as chips and crisps, whereas in much of the rest of the world the same products are referred to, respectively, as french fries and chips. In this book, which is aimed primarily at a British readership, the British terms will be used.

The co-products of such processing activities have a number of advantages over the stockfeed supplies that were previously available to livestock farmers. The potatoes are grown on selected fields under strictly defined conditions. Pesticide and herbicide treatments are restricted to approved products, and their application by trained personnel must conform with the guidelines set out in the "Green Book" (The UK Pesticide Guide; Anon, 1999). Selected crops are rigorously tested for disease and laboratory tests ensure that the potatoes conform to exacting limits on pesticide residues. It would be entirely appropriate to regard them as potatoes of choice, rather than the discards of any selection system.

All processors begin their activities by subjecting the selected potatoes to a cleaning regime that often includes water flumes in which the potatoes rub against one another as they are carried along in the flow. A series of brushes is then employed to scrub the tubers as they are tumbled in powerful water sprays. After washing, the potatoes are subjected to various procedures depending on the objectives of the factory, and collectively these result in a range of sliced, chipped, diced and other potato products - many of which are at least partially cooked. It is axiomatic that washed potato must never come into contact again with the muddy water streams separated at the start of processing, and this requirement applies just as surely to the co-product fractions as to the selected products that will be marketed directly to the human food market.

Frozen and chilled potato processing

From a standing start in the 1960's, the frozen potato industry has grown to become the principal market for the British potato grower. During the last decade, the volume of potatoes processed into chips (french fries) has almost doubled, and the quantity continues to rise. The range of frozen products has also widened and supplementary lines have been introduced in the factories in order to increase the efficiency of potato processing. However, the tighter the specification of the core products, the greater the proportion that fails to meet the criteria. The production of feed grade co-products is thus an inevitable consequence of higher standards within the processing factory.

Frozen chips represent the principal market for the British potato grower
(courtesy: McCain Foods GB Ltd)

However, the first operations within the factory remove fractions that are neither food nor feed. Soil and stones and any extraneous vegetable matter are separated in the initial phase, and any hollow or diseased potatoes are floated off. The vegetable wastes are carried to landfill sites, while the soil is collected in mud pits and later spread onto arable land. An exception is often made for soil collected from imported potatoes, much of which is carried to landfill in order to protect the British grower from any imported soil-borne diseases.

The cleaned potatoes are then peeled, usually by a pressurised steam process, in preference to the simple abrasive peeling that is commonly used by the crisp manufacturers. Application of steam at a pressure of up to 15 atmospheres for a period of 15-30 seconds is followed by a rapid release of pressure, and this loosens the attachment of the outer layers to the inner potato flesh. These layers are then abraded as the potatoes are tumbled in a rotating drum. For non-ruminant animals there is a critical distinction between the two types of peel, in that steam peeling results in a cooked co-product which is typically of much higher nutritional value than raw potato. The cooked co-product is often referred to as **steam peel** or **potato feed**. The peeling conditions are closely controlled and the depth of peel is related to the age of the potato. The thin skins of new potatoes are separated in a relatively short time, whilst older potatoes are subjected to longer steaming and deeper peeling. From the animal feed viewpoint, the peel from older potatoes is preferred because it contains more of the potato flesh and is consequently richer in starch and lower in fibre content.

Peeled potatoes are moved into the factory where any defects and partially unpeeled potatoes are removed. These off-cuts, together with the slivers cut from the outside edge of potatoes, are directed towards the animal feed market, where they are known variously as **nubbins, hopper or slice**. Although the potatoes were steam-peeled, the inner flesh that provides this feed fraction is essentially uncooked.

Selected, peeled potatoes are then ready for cutting into chips, which may take a variety of forms.; viz: long or short, fat or thin, straight or crinkle cut, in order to satisfy the various markets for deep fat frying, conventional oven or microwave chips. The various specifications have implications for the nutritional value of the co-products - the greater the surface area of the chip the higher the proportion of oil that will be absorbed during cooking, and the richer the oil content the higher the energy value of the feed. For any specific oil content, the nutritional value is directly related to the dry matter level of the chips.

Whichever shape the potatoes are cut, there are inevitable losses as the knives release the starch-rich cell contents into the water stream that carries the potatoes around the factory. Within the European Union, it has been recognised that this is a "process water stream" that carries food quality materials through the factory and, at the end of the process, the starchy liquor can be used as a source of animal feeds. The term process water distinguishes it from "waste waters" that are not permitted to yield feed co-products, and any contamination of process water, with soil or other effluents, downgrades the stream to waste water with irrevocable consequences (EU, 2000).

After cutting, the raw chips are blanched by immersion in hot water for several minutes. During blanching, the raw chips are subjected to water temperatures of 60-80°C and, since the gelatinisation of potato starch occurs at temperatures above 55°C, all the potato products and co-products from this point in the process can be considered to be at least partially cooked. Blanching reduces the sugar level of the potato and this has important implications for the colour development during frying. Although the colour of the chips is of minor importance to the animal feed market, the blanching process is useful in that it inactivates enzymes which may promote decay during storage on the farm (Wilmot, 1988). Blanching also reduces the effectiveness of the chymotrypsin inhibitor, present in raw potato, that impairs protein digestion in monogastric species (Whittemore *et al.*, 1975). The blanched potatoes are then partially dried, a process that removes a variable amount of water of between 5 and 25 per cent. Such drying enhances the nutritive value of the fresh material in proportion to the extent of water loss.

Some of the potato pieces leaving the cutters are unsuitable for the production of the required type of chip, and they are diverted into a separate line to be cooked, mashed and dried into potato flake. This is a tightly-defined operation where overcooking spoils the texture of the finished product and undercooking leads to a high rejection of mash by the steam-heated rollers. The task is made more difficult by the innate variation that typically occurs within each batch; a wide range of moisture content implies a need for a wide range of cooking times. Inevitably there is some product that fails to meet the exacting standards of the process, and this non-spec material becomes the **potato mash** that is available as animal feed.

The partially-dried, blanched product that is suitable for frying follows a different route. Frying is a carefully controlled process that can take between 30 and 120 seconds in oil maintained at a temperature of between 165-195°C. The precise cooking conditions are defined by the type of chips that are being made. Oven chips, and particularly micro-wave oven chips, need a longer period in the frier; chips that will be deep fried before consumption need less time in the factory frier. As noted earlier, the precise specification of the end product, which includes length, colour and consistent performance in the kitchen, inevitably means that a proportion will always be rejected and these **chips** become available as animal feed.

Differences in cooking times in the frier lead to differences in the nutritive value of the co-product chips – in general, the longer the frying time the higher the dry matter content of the chips. Consequently, the narrow chips that are intended for subsequent oven preparation are of significantly higher feeding value than the broad, straight chips designed for further frying – such "oven chips" tend to be higher in both oil and dry matter content. At the present time, limited storage facilities at the factories do not permit separate marketing of the different types of co-product chip, and the best advice that can be offered to the livestock farmer is an indication of the likely proportions in each load. As processing facilities expand, and the opportunity for niche marketing a well-defined co-product develops, some of the precision available in the human market may be transferrable to animal feed sales.

The oil in which the chips are fried is specially selected to ensure the required frying temperatures, and with particular regard to the prevailing attitude of public health specialists towards dietary fat (see NACNE, 1983; COMA, 1991; DOH, 1994). The oil content and its fatty acid composition also have implications for animal nutrition. Oil is a high-energy source that can boost the value of any diet, and this is of particular value when feed intake is restricted by high temperatures or early lactation. For the sow, oil can have specific benefits on piglet survival and weaning weights, but an excessive supply can lead to overfat pigs. Ruminant digestion can also be depressed by an excessive consumption of oil, particularly when it contains a significant concentration of polyunsaturated fatty acids. In the British frozen potato industry, chips are usually cooked in a blend that may include sunflower, rapeseed and palm oils. Frying in palm oil may be of particular benefit where chips are to be used for ruminant animals, because its relatively saturated nature may enable higher inclusion rates of chips in the diet. Irrespective of the origin of the oil, providing it is well defined, the dairy farmer may be able to use chips to manipulate the composition of the milk in a predictable manner. That would open up the prospect of a niche market for chips, and could potentially place a premium value on the co-product.

At the end of the production line, the process waters are transferred to a primary clarifier where the starchy solids are allowed to settle. Entry to the clarifier is protected by a 1mm screen that excludes larger potato pieces, and these are commonly collected and added to the offcuts and slivers obtained previously. The starchy sediment that separates from the process liquor is removed and centrifuged to increase its dry matter content. The centrifuged solid fraction is marketed to the pig industry as a high-energy liquid feed under the name prime potato puree or, as it is more commonly known, PPP.

Co-products of frozen and chilled potato processing

Potato feed (steam peel)

Potato feed comprises the outer layers of potato, removed by steam peeling during the initial processing stage. The material is freed from attachment to the potato flesh by a sudden drop in pressure within the steam chamber, and sloughed off by abrading rollers. In some countries potato feed is referred to as "steam peel", to indicate that the material is a cooked product, but "peel" seems an inadequate description for a product that comprises a significant proportion of the starchy potato centre.

Much of the potato protein and fibre is contained within these outer layers, but the starch content is also substantial. "Older" potatoes are peeled more deeply than "new" potatoes and the resulting potato feed has a starch content that approaches the level found in oat grain. With significant protein, fibre and starch contents, potato feed is a well-balanced feed material for cattle and sheep, and a useful though fibrous ingredient of pig rations. In some pig feeding systems, where a high proportion of refined food co-products is used, potato feed is selected because of its fibre content, in order to boost the fibre level of the whole diet, although it is typically macerated to improve its pumpability.

Potato feed has a moisture content that is higher than that of the original potato, owing to admixture with steam during peeling. Much of this water becomes chemically bound as the starch granules gelatinise, and the co-product presents itself as a viscous, porridge-like slurry that is best held in special storage tanks. From such tanks, the material can be pumped to a mixing plant and thence by pipeline to fattening pigs. For ruminant animals, the co-product can be delivered from the tank to a mixer wagon and thence to feeding troughs as part of a complete ration. Potato feed can also be stored in an open (bunker) silo where, within days, it forms a firm gel and this prevents the product from flowing. However, in the absence of a retaining wall, the co-product settles to a relatively thin layer and it is uneconomical in the use of silo space. If stored in this way, the co-product can be handled by tractor bucket and either fed on its own or incorporated into a mix as described above.

The gelling properties of potato feed have prompted a novel use for the co-product, as a sealant for silos filled with other moist feeds, which may be prone to aerobic decay unless well protected. Brewers' grains, sugar beet pulp, grass and maize silages have all been sealed successfully by a capping layer of potato feed, which

Potato feed is a well-balanced feed for dairy cows
(courtesy: James & Son (Australia) Pty Ltd)

flows into all the air spaces within the surface layer before forming a gel that will exclude air for a number of months. Not only does it provide an effective seal of edible potato, capping with potato feed brings additional labour-saving benefits. The mechanical application of the co-product saves labour when the silo is being sealed and, because it replaces both the sheet and the need for additional surface weighting, when opened for feeding the potato capping saves the effort that may otherwise be spent moving bales, tyres or sand-bags. However, potato feed capping is a good idea that may have been partially overtaken by other events. Although its advantages in terms of the effective sealing of the underlying feed are not in question, potato feed capping does not comply with Codes of Practice for the safe storage of moist feeds (BFBi, 2000; NDFAS, 1998b). Both these codes contain a blanket recommendation that the surface of moist feed stores – including any overlying potato layer - should be sheeted to prevent contamination by birds, cats and vermin. Thus the potato would itself need to be covered in order to to avoid the risk of contamination.

For ruminant animals, potato feed is a palatable feed material that is readily consumed and rapidly digested by both cattle and sheep. Its nutritional value has been determined in vivo at the ADAS Feed Evaluation Unit (ADAS, 1994), where the digestibility of protein, starch and gross energy was found to be high. Further research at SAC Aberdeen found that potato feed increased the rate of forage digestion, thus raising the possibility that the feed may be expected to increase feed intake. Potato feed has comprised as much as 50 per cent of the diet in experimental studies, but a lower limit is likely to be more appropriate for practical rations (Rooke *et al.*, 1997).

Dairy cows typically consume up to 25-30 kg potato feed per day, although larger quantites have been used (Wilkinson and Kendall, 1997). This amount provides sufficient energy to replace 3-3.5 kg of cereals and adds extra protein to stimulate microbial synthesis in the rumen. Dairy heifers can be reared on diets containing a high proportion of potato feed combined with a little hay or straw; for such stock, 15 kg potato feed could replace 2 kg rearing nuts. Increasing amounts of potato feed can be fed to beef cattle as the animals grow and, in the later stages of fattening, 40 kg of potato feed can replace up to 5kg cereal grain. For reasons of practicality, sheep are rarely given the opportunity to consume potato feed but, when it has been offered to them – as in the metabolism studies of Rooke *et al.*, (1997) – the feed was readily consumed. Potato feed may be particularly useful as part of a change-over ration for lambs which are being brought from pasture onto an intensive finisher ration.

Dairy cows typically consume 25-30 kg of potato feed per day
(courtesy: James & Son (Australia) Pty Ltd)

For all ruminants, potato feed is best regarded as a concentrate feed which needs to be fed in a balanced ration, with an adequate proportion of fibrous feeds in order to ensure optimum conditions in the rumen.

For pigs, potato feed can play a valuable role in the diets of both sows and growing stock. The feed provides more protein than the cereals it commonly displaces, and roughly twice the amount of lysine (on a dry matter basis). Research at SAC Aberdeen (Edwards, 1993) has shown that all parts of potato feed are highly digestible, and together they add up to a useful DE content of 14 MJ/kg DM. Potato feed can comprise up to 20-25 per cent of the diet for growing and finishing pigs, and even more can be fed to dry sows (Edwards and Livingstone, 1990), which have a greater capacity to cope with a bulky diet.

Further processing of potato feed: In the form in which it exits the steam peelers, potato feed can present physical handling problems for the farmer; it is too fluid to be stored efficiently in a bunker silo, but may be too viscous to be pumped satisfactorily through relatively narrow pipelines. Centrifuge trials aimed at producing a solid product of much higher dry matter content have failed on two counts. The separated centrate was neither water nor liquid feed, but a thin, low dry matter, “milky” liquid with a BOD value high enough to represent an unwelcome burden on any effluent processing facilities. Notwithstanding the removal of this centrate, the dry matter content of the solid fraction remained relatively low, and its fibre content had been increased (James & Son, unpublished results). Consequently, the centrifuged solids represented a fraction with poorer nutritional value than the original potato feed, and appeared to offer little compensation by way of a substantially reduced transport cost (per unit of dry matter).

Pressing may represent a more useful choice for further processing. As with centrifugation, two fractions are obtained; one liquid and the other solid, but, compared to the centrifuged material, the liquid fraction contains a much larger proportion of the potato feed solids and could potentially be regarded as a feed material. Both the proportion of liquid permeate, and its composition, depend on the size of the screen through which this fraction is pressed. However, it typically contain much less of the potato fibre, though its dry matter and protein contents are only marginally different from the ingoing potato feed. **Potato feed permeate** is a highly digestible liquid that could be offered as an easy-to-use potato co-product feed for pigs. The **potato feed solids** are a more fibrous, and less digestible fraction, but they are marginally drier than the original potato feed and they appear to be more suitable for storing in a bunker silo for ruminant feeding.

Potato off-cuts (potato hopper) and canning potatoes

Potato hopper is the name popularly given to the slivers and off-cuts of uncooked potato that are considered unsuitable for the production of chips or other processing.

In essence, the co-product comprises small pieces of raw, peeled potato and contains a lower level of protein, fibre and minerals but a slightly greater proportion of starch than whole potatoes. Small canning potatoes rejected in the factory because of minor blemishes are of similar composition and utility. As a peeled material, it is unquestionably cleaner than stockfeed potatoes that may be considered as an alternative purchase, and its chopped nature makes it a safer feed than whole potatoes since there is much less danger of choking. On occasion, however, whole peeled potatoes find their way into the potato hopper, which precludes any claim to zero risk. This potential problem has been addressed in some factories, where the co-product material is subjected to further chopping in an attempt to eliminate whatever risk these tubers may represent.

Uncooked potato tends to discolour quickly, but the co-product remains a palatable, succulent feed capable of supporting a high level of animal production. Because of the presence of an anti-nutritional factor in raw potato that interferes with protein digestion in non-ruminant animals, potato hopper is not usually regarded as a suitable feed for pigs, although Edwards and Livingstone (1990) suggested that it may be fed to "dry sows and finishing swine". It is highly regarded by ruminant farmers and is particularly suitable in beef fattening rations. In common with most potato co-products, potato hopper is highly digestible – Stanhope *et al.* (1980) found potato starch digestibility by beef cattle to be 99.1 per cent, and potato hopper comprises around 75 per cent starch. Since these authors also found that more than 94 per cent of the starch was digested in the rumen, it is clearly important to ensure that this highly concentrated feed is adequately balanced with a regular supply of forage. Consumption of fibrous feed is needed to stimulate saliva production and provide an alkaline medium that will neutralise the acidity that develops as a result of the rapid digestion of potato starch.

Dairy cows can be fed up to 15 kg potato hopper per day in place of about 3.5 kg of rolled cereals, dried sugar beet pulp or other energy feeds. The feed will tend to have a positive effect on dry matter intake. Dairy heifers find the co-product attractive and this makes the feed suitable for use in conjunction with less palatable feed materials, such as straw, that are commonly reserved for this class of animal. A daily ration of 10 kg potato hopper may need to be supplemented with a protein balancer, such as brewers' grains, but the hopper substantially reduces the heifer's need for other compound feed supplements. Beef cattle can be fed liberal quantities of potato hopper according to animal size and the required level of performance. Rations comprising 60 per cent of the dry matter in the form of raw potato were fed in ADAS trials (Edwards, 1982), and growth rates approaching 1.5 kg per day were recorded for both bulls and steers. Sheep are not commonly offered potato hopper, but it is potentially a high value feed that could support a good level of performance – Archer *et al.* (1980) concluded that raw potato was a feasible alternative to grain for use in feedlot diets for lambs. Because of its unfamiliarity to sheep, it would be essential to feed potato hopper with caution, introducing it gradually and ensuring that it is fed in conjunction – preferably mixed - with a fibrous feed.

N.B. Potato hopper must be adequately drained before leaving the factory. The presence of a significant proportion of process water rapidly leads to the development of an unpleasant smell, and both farmers and their stock find this unattractive. Even where the extraneous water amounts to no more than five percent of the total volume, the appearance may be of a feed "standing in water", and the reputation of the co-product as a valuable succulent is damaged. On the farm, this water runs freely from the stored co-product, and carries with it a proportion of the potato starch. To the farmer's eye, it represents a loss of weight, a loss of feeding value and a pollution threat. Process water can be handled more safely and more easily on the factory site, and its separation from the co-product at that point dramatically enhances the perceived quality.

Potato off-cut /steam peel skins mix (potato slice)

At some potato factories, the off-cuts and slivers are mixed with peel that has escaped the initial peeling process and been detached subsequently. Before being mixed with the raw potato off-cuts, this peel has travelled round the factory and it is collected from the process water as it enters the primary clarifier. As this stream moves through the factory, the inner fleshy layers become detached from the potato skin and its feeding value is consequently reduced. Thus the addition of this material to the slivers and off-cuts yields a lower value co-product. This combined material is sometimes marketed under the name potato slice.

Potato slice is suitable for ruminant animals. It is an energy feed containing a substantial starch content in conjunction with a significant fibre level. Because of its starch content, potato slice needs to be introduced to the diet gradually, but it is less likely to lead to problems of acidosis than the higher starch containing potato hopper.

Dairy cows can be fed up to 15 kg potato slice per day in place of about 2.2 kg of rolled cereals, dried sugar beet pulp or other energy feeds. Dairy heifers can be offered 10 kg potato slice in place of 1.5 kg cereal grain. Beef cattle can be fed increasing amounts of potato slice as the animals grow and, as the stock reach marketing weight the diet may include as much as 18 kg potato slice per head per day. Sheep may also be offered potato slice but the product should always be fed with caution. Selective feeding may result in some animals consuming a diet that is largely composed of potato off-cuts and little peel. Thus, even though it is a more fibrous co-product than potato hopper, this feed should still be introduced with care and fed in conjunction with adequate amounts of hay or palatable silage.

Potato mash

Potato mash comprises cooked, peeled potato that forms a staple constituent of the human diet and represents a high value feed suitable for cattle, sheep and pigs. It is similar in composition to the slivers and off-cuts that are marketed as potato hopper, but cooking gelatinises the potato starch and increases the rate of digestion by rumen microbes. Laboratory studies with rumen liquor have shown that cooked potato is digested much more quickly than even the most rapidly digested cereal grain (Beever *et al.*, 1999). The cooking also inactivates the anti-nutritional factor that makes raw potato an unsuitable feed for pigs. Where it can be incorporated easily into the ration, potato mash is an attractive feed for pigs – more digestible and richer in lysine than the wheat it could easily replace.

The high rate of digestion must be borne in mind when offering potato mash to ruminants, since excessive consumption can quickly lead to acidosis. The co-product should always be fed in a mixed ration with adequate amounts of fibrous feeds, unless it is to be used in strictly restricted quantities. In the latter situation it may be prudent to consider splitting the feed between different meals. A combination of potato mash with rapidly available protein sources may stimulate microbial protein production.

Dairy cows can be fed up to 25 kg potato mash per day in a mixed ration that will be offered to appetite, and its rich starch content may be expected to boost milk protein output. Dairy heifers would need to be fed a more restricted amount of potato mash; with average quality grass silage, no more than 6-7 kg potato mash would be needed to sustain an adequate growth rate. Beef cattle, fed to appetite on a complete ration, can utilise large quantities of potato mash, and this is probably the most appropriate use for this energy-rich feed. Intensive rations could include up to 40 per cent potato mash, on a dry matter basis, providing that it is continuously available and balanced with an adequate proportion of dietary fibre. Sheep have rarely been offered potato mash and, although it is a highly nutritious ingredient, it would need to be fed as part of a well-mixed fibrous ration if its potential as a high value feed is to be realised without running the risk of excessive consumption by individuals. A prudent inclusion rate of about 10 per cent of the dry matter of a well-mixed ration may be advisable.

Potato chips

Potato chips are the co-product of frying thin sticks of partially-cooked, peeled potato in oil. In nutritional terms, the feed is a starch-rich material whose high energy level has been further boosted by the absorption of oil. With the exception of beer, potato crisps, flour, confectionery and ice cream, chips represent the highest energy feed source outside the range of feed-grade oils and fats. The energy content varies between different types of chip; narrow chips with a relatively high surface area are typically richer in oil, and those designed for final preparation in an oven (Oven Chips) tend to have a lower moisture content as a result of longer frying in the factory.

Potato chips have an energy level higher than almost any animal feed
(courtesy: James & Son (Grain Merchants) Ltd)

This co-product is suitable for feeding to cattle, sheep and pigs, though consideration must always be given to the potential impact of the oil. Chips cooked in rapeseed and sunflower oils contain a higher level of polyunsaturated fatty acids than those cooked in palm oil and, when fed to excess, such acids are known to depress microbial activity and consequently fibre digestion in the rumen. But the oil confers rumen benefits too, suppressing energy wastage in the form of methane gas. Rumen digestion of cereal grain typically wastes about ten per cent of the gross energy by conversion to methane (MAFF, 1990), whereas digestion of the higher energy value potato chips results in the loss of only three per cent of that energy in gaseous form (Rooke *et al.*, 1997). For pigs, the oil can have specific benefits, in addition to its high energy value, when it is fed to the sow just before and just after she has given birth (Pettigrew and Moser, 1991; Holness and Mandisodza, 1985). However, because of its highly efficient use in fat synthesis, dietary oil for the fattening pig can lead to an undesirably fat carcase. The oil intake of the pig needs to be balanced by an appropriate intake of available amino acids.

Dairy cows can be fed approximately 5 kg of chips per day, with the optimum feeding level being determined by the oil supply from other feeds. This amount of chips could replace 2.2 kg of rolled barley and would result in an increase in the energy density of the ration. Dairy heifers typically require little in the way of concentrated feed but, where available, chips can replace cereals or sugar beet pulp at an approximate ratio of two kilogrammes of chips to one of dry feed, with a maximum allowance of about 1 per cent of body weight. Beef cattle can make the greatest use of chips when offered to appetite as part of a mixed diet in an intensive system. High growth rates in excess of 1.9 kg per day, with a feed conversion of only 4.3 kg feed per kg of gain, have been achieved with complete rations containing as much as 40 per cent chips on a dry matter basis (Thomas, 1996). Few sheep are given the opportunity to perform on a diet containing potato chips, but metabolism studies (Rooke *et al.*, 1997) and anecdotal evidence from the field confirm the palatability of the co-product and highlight its rich potential. Lambs were reported to have run along the backs of their mothers and flung themselves into the tractor bucket delivering a chip-based moist mix.

Sows around farrowing should ideally be fed a diet containing about eight per cent oil (Crawshaw, 1994), and this may permit a substantial inclusion of potato chips. Such a diet would promote piglet viability and an increase in growth rate, leading to a greater weight of weaned pigs (Moser *et al.*, 1978; Pettigrew, 1981; Seerley, 1984). Growing and fattening pigs typically respond positively to dietary fat levels of up to five per cent, which may permit the inclusion of a modest quantity of potato chips where available. Beyond that level, animal performance tends to be moderated by declining feed intake, although the commercial benefit may continue to improve with increasing feed conversion efficiency (Pettigrew and Moser, 1991).

High ambient temperature may be a problem in certain seasons with animal performance restricted by a reduced intake. In such conditions, potato chips as a palatable, oil-rich feed may be of particular benefit – a combination of high energy density and greatly improved energy utilisation within the tissues make this feed an almost unbeatable choice for this niche market.

Other fried products

In recent times, potato chip manufacturers have begun to extend the range of products that they offer. Such developments are likely to continue and it may be expected that a succession of new products will replace or add to those already on the market. Inevitably, a proportion of them will become available to the animal feed market. Their value will depend predominantly on the presence or absence of peel, the loss of dry matter during processing, and the proportion of absorbed oil. Some examples are described below.

Jacket Wedges: This material comprises unpeeled potato, cut into large wedge-shaped pieces, which are blanched and then fried. The typical weight of the wedge is some four times that of an oven chip but the surface area is relatively smaller, with implications for the amount of oil that will be absorbed during cooking. The oil content of jacket wedges, at about 12 per cent of the dry matter, is lower than would typically be found in more conventional chips. The presence of peel may increase the protein and mineral contents, though only marginally. This product is lightly seasoned, which lifts the sodium content to approximately 0.5 per cent of the dry matter – a level that should pose no problem in the diet of ruminants or even pigs.

Southern Fries: This product is a typical oven chip that has been dipped before frying into a seasoned batter. The batter may be potato or wheat-starch based, and include relatively mild seasoning with salt, pepper and garlic. The small dimensions of these chips lead to a relatively high oil content of around 20 per cent in the fried product – a typical level for oven chips. Cooking results in an unusually high dry matter content, close to 40 per cent and the seasoning also boosts the sodium content to a level of around 0.75 per cent of the dry matter. This would pose no problem for ruminants but the co-product would need to be rationed carefully if fed to pigs in conjunction with other feedingstuffs that are rich in soluble minerals.

Hash Browns: This product takes the form of shredded and well-seasoned potato patties, and in chemical composition they closely resemble potato chips. In the factory the potatoes are chipped in the normal way, before being shredded and then formed into large patties. Because of their roughened exterior the surface area of the patties is relatively high and the product absorbs an appreciable amount of oil. Analysis has revealed an oil content of 19 per cent of the dry matter, possibly higher than the product specification because of a greater proportion of broken and edge pieces in the co-product fraction. This material also has a high dry matter content, at around 38 per cent and, significantly, a high sodium level of 1.75 per cent of the dry matter. The sodium level may be quite variable depending on the proportion of broken and outer material and the co-product should pose no problem for ruminant animals, but it would not generally be recommended for pigs.

Individually, these co-products are of minor significance, but collectively their volume may be expected to grow. At present, they are available only in a mixture with potato chips where their nutritional value may be regarded as broadly similar. However, it may be expected that factories will manufacture these alternative products in batches and, at those times, the co-product output will reflect a relatively high proportion of them.

Prime potato puree and potato starch

Prime potato puree (PPP) is the starch-rich feed material recovered at the end of the line from the process waters that carry potato products around the factory. PPP represents fine fragments of potato lost at various stages of the processing, and particularly at the point where the potatoes are sliced, chipped or diced releasing the starch-rich content of damaged cells into the water stream. Prior to the isolation of PPP, in fact immediately after the potatoes have been cut, ungelatinised potato starch is recovered from cold water and this material is usually marketed to the food and packaging industries. Only occasionally does this uncooked potato starch become available as an animal feed, but laboratory studies have shown it to be a potentially useful slow-release starch source for ruminant animals (Beever *et al.*, 1999).

The starchy sediment that remains in the process waters is later subjected to treatment with hot water and this brings about a degree of starch gelatinisation. Although this renders the starch unsuitable for some other purposes, gelatinisation tends to improve the nutritive value of the material for farm animals, marginally increasing its digestibility but exerting a much greater effect on the rate of digestion. In addition to the starch-rich potato flesh, this sediment also includes a proportion of peel that becomes separated after the initial peeling process. The inclusion of this peel increases the fibre and protein content of the material that is eventually separated as PPP so that, although rich in starch, this co-product is significantly different from the almost pure potato starch recovered earlier. The proportion of peel varies between different factories so that it is advisable to obtain the relevant PPP specification before formulating rations.

PPP can most accurately be described as partially-cooked, and when the co-product first became available this led some advisers to question its suitability as a pig feed, since uncooked potato is not usually recommended. Their concerns included the ability of the pig's enzymes to digest ungelatinised potato starch, the negative impact of the anti-nutritional factor in raw potato and the possible encouragement of an undesirable type of starch fermentation in the hind gut. Tests for anti-trypsin activity in PPP have found only low levels (James & Son, unpublished report), and this finding seems consistent with the description of the anti-nutritional factor as a heat-labile, water-soluble component (Whittemore *et al.,* 1975). Before separation from the process waters, PPP has been subjected to substantial volumes of water at different temperatures, and it seems likely that such processing has either inactivated or removed much of this unwanted feature. Farm experience has confirmed that PPP can be used efficiently for both growing and fattening pigs, and a most illuminating investigation of the site of digestion was carried out at SAC Aberdeen (Rooke, 1999). This study found that PPP may be used for weaner pigs of only 10-20 kg liveweight; even with such young pigs, more than 95 per cent of the starch was digested, and much of this digestion was effected by enzyme activity in the small intestine.

After separation from the process waters, PPP is centrifuged to increase its dry matter content and the ultimate dry matter level is broadly under the control of the operator. The profitability of moist feed marketing is strongly linked to dry matter content but, if the dewatering of PPP is taken too far, it can transform a pumpable fluid into a solid cake that would create considerable difficulty for pipeline-fed pig farms. The optimum viscosity for liquid feeding occurs when the PPP is in the region of 17-23 per cent dry matter, and the majority of the supply falls within this range. However, if the co-product were to be used as a ruminant feed, a higher dry matter content may be desirable, since this would allow the material to be transported more economically in a bulk vehicle in place of a tanker.

On the farm, PPP is an alternative to cereal grain in the diet, without the need for further processing before feeding. It is most commonly stored in a dedicated storage tank and used as a liquid feed in diets for fattening pigs, being typically included at 15-20 per cent of the ration dry matter. The co-product is recognised by pig farmers as a valuable energy source that has commendably little tendency to separate into different layers during storage. Their principal criticism of the co-product is the variability in the dry matter content of different consignments. This is a problem largely caused by variable throughput at the centrifuge, and should be capable of resolution. Appropriate control at the factory would improve the image of the co-product and increase the farmer's ability to feed his pigs a consistent ration.

PPP has not been widely used for ruminant animals, but this starch-rich co-product has the potential to provide a high-energy supplement that would be suitable for inclusion in rations for the most highly productive stock. Its use for ruminants should be advised with caution in order to avoid any risk of acidosis – PPP has a similar starch content to wheat grain, but its digestion in the rumen may be much more rapid. Laboratory studies with rumen liquor have shown that the degradation of PPP is both faster and more extensive than that of wheat (Beever, 1999). Care must thus be taken to ensure an adequate provision of dietary fibre for the stock, and this would best be achieved by the incorporation of PPP into a complete, mixed ration.

Potato puree feed

Potato puree may also be available in combination with potato feed, as at the Garden Isle factory in Wisbech, and this combined co-product is known as potato puree feed (PPF). The puree removed from the process waters is centrifuged before blending with the steam peel that was removed at the start of the processing chain. Although there is typically no attempt at the factory to control the proportions of each component, experience has shown that the composition of PPF is relatively consistent. Because of the way the two materials are jointly drip-fed into the factory silo, the combined co-product is well mixed and there appears to be little tendency for the two materials to separate.

As a feed, PPF combines the virtues of its two components; it contains more starch than would be typical of potato feed and more protein and fibre than most sources of PPP. Since potato feed is a cooked material and PPP is partially cooked, the combined co-product may be regarded as mainly cooked, and it is suitable as a feed for both ruminant animals and pigs.

The dry matter content of PPF is typically 15-16 per cent and it presents itself as a viscous porridge-like slurry. As such it is best stored in a sealed silo and, since these are more commonly found on pig farms, this co-product blend is usually fed to pigs. For fattening pigs, PPF typically forms 15-20 per cent of the ration dry matter and more can be used in dry sow diets.

PPF is also used on ruminant farms, principally as an ingredient of a complete ration for beef cattle. With an ME value close to 12.5 MJ/kg dry matter, and a starch content in excess of 40 per cent, this co-product blend is an ideal alternative to cereal grain. PPF adds succulence to the mix, and aids the distribution and retention of minerals and other dusty materials that may have a tendency to separate.

Potato crisp production

Crisped potatoes have been known since the 1850's, but during the next hundred years the very limited shelf-life of the fresh product restricted its availability. In Britain before the Second World War, crisps were mainly produced as a holiday snack in sea-side towns, and they were typically consumed on the day they were produced. The development of sealed bags in the 1950's provided the breakthrough that permitted crisp production on a large-scale, and the industry expanded at a prodigious rate. From a novelty item sampled on holiday, crisps have become Britain's most popular savoury snack, and more than eight billion (30g) packets are purchased each year.

In essence, modern crisp production still resembles the process carried out in the early days; peeled potatoes are sliced into wafer-thin sections and cooked in oil until a light-golden brown. But the process is now more closely controlled in order to produce a uniform product. In particular, the sugar level of the potatoes is monitored carefully and finely adjusted to avoid significant variation in colour, and especially the development of the less desirable dark-brown crisps.

Crisp manufacturers prefer to peel their potatoes by abrasion, rather than the steam-peeling typically chosen by frozen chip producers. Abrasion is considered to remove a smaller peel fraction, and to yield a product in which the turgor of the cells is undiminished and more suitable for subsequent operations. The implications for the co-product market are that abrasive peeling yields a smaller volume, the abraded peel contains less of the potato flesh and the whole of the peeled fraction remains completely uncooked. In recent years, further reduction of the peel volume has been achieved by an increasing use of freshly harvested potatoes in place of older potatoes removed from store. The impact of this change on co-product volume can be gauged from the reduction in peeling times, which have been lowered from a range of 30 to 60 seconds down to a typical period for new potatoes of only 10 seconds.

After peeling the potatoes are checked for defects, and off-cuts of peeled, uncooked potato become available as animal feed. Selected potatoes are transferred into centrifugal slicing machines that convert whole tubers into slices about twelve cells thick. The blades themselves are about two cells wide and the slicing action damages the layer of cells on each side. Thus the consequence of slicing is an approximately 18 per cent loss of potato into the process water. Potentially, this loss could become a substantial source of a starch-rich co-product feed, but it is not currently supplied to the animal feed market. The starch is recovered from the waters, pressed to some 50-60 per cent dry matter and sold to the packaging industry. In addition to the slicing loss, there is an inevitable rejection of "nubbins", the small rounded end pieces that are too small to be used as crisps and, in conjunction with the off-cuts separated earlier, this material does become available as animal feed. Since no heat has been employed in the processing, such materials remain uncooked and they are consequently more suitable for cattle and sheep rather than the non-ruminant market.

The potato slices are washed in cold water to remove sugars and any free starch and they are then ready for the frying operation. Cooking is typically a continuous process in which steam is driven off and some of the oil absorbed. Because the slices are so thin, water removal is almost complete and the cooked product has a dry matter content in excess of 98 per cent. The relatively large surface area also leads to significant oil uptake and the cooked crisps comprise some 34-35 per cent oil. Some of the crisps are rejected at this stage because of spillage or unsatisfactory colour and this high value material becomes available as a (cooked) animal feed.

Palm olein is the almost universal choice of cooking oil for the snack food industry. It is the liquid fraction of palm oil and has been used as a cooking oil for thousands of years. Its fatty acid composition comprises approximately 45 per cent saturates, 42 per cent monounsaturates and 12 per cent polyunsaturates. The high oil uptake by the crisps necessitates continual replenishment of the system with fresh oil and, together with the thick steam blanket that overlies the friers and protects the oil from oxidation, this continual turn-over means that only a small proportion of oil needs to be discarded. The limited supply of discarded oil, which would be fully traceable to the factory, could be used directly as an animal feed, but this is not its usual destination. Typically, the discarded oil is returned to the refiner for further processing and, following that, some of it may become available to the animal feed market, possibly in blended form. However, following the serious and far-reaching dioxin problem that arose in Belgium as a result of the adulteration of recovered vegetable oil, new legislation will impose strict controls on the re-use of cooking oils as animal feeds. The practice will continue to be acceptable provided that HACCP plans have been instituted and full traceability can be demonstrated (EU, 2001).

Crisps are available in Britain with a wide range of flavours. The industry prefers to use the term "seasoning" rather than "flavouring", and it has been engaged in this practice since the early days when salt was included in a twist of paper in each bag. Modern technology applies the seasoning within the factory, though "ready salted" is still the most popular variety. The seasonings are applied at an application rate of approximately six per cent and at this rate the sodium content may have significant implications for the animal feed industry, particularly if the co-product were to be offered to pigs. The additional use of other sodium salts, such as sodium acetate and diacetate to impart a vinegar flavour, and sodium glutamate as a flavour enhancer, is of further note. An average sodium level of 0.8 per cent is regarded as typical of seasoned crisps, although co-product crisps may be expected to have a lower sodium level owing to their inclusion of crisps rejected prior to the seasoning process.

Co-products of potato crisp production

Abraded peel

Abraded peel is the thin, outer layer of the potato that has been physically scraped from the inner flesh. Peeling times are typically short, and the peel contains much less of the starchy flesh than the steam peeled co-product. Its composition is quite unlike the rest of the potato, comprising a very fibrous feed material with a low digestibility (cf the enzyme digestibility, NCGD, of abraded peel is 37.5 per cent, substantially lower than the typical 80 per cent value for steam peel). The peel is removed under a light stream of water but it is subsequently transported in water and the separated co-product can have a very low dry matter content (< 10 per cent). However, the ungelatinised starch does not swell in cold water and the associated water is not chemically bound. This lack of binding facilitates the subsequent removal of the water, offering the prospect of a substantially higher dry matter content if the peel can be pressed or centrifuged. Higher dry matter levels often translate into higher value feeds and efforts in this direction are already being made at some factories.

Abraded peel is not subjected to any heat treatment and it consequently remains an uncooked product, more suited to cattle and sheep than non-ruminant animals. However, the fibrous, indigestible nature of abraded peel would render the co-product unsuitable for pigs even if it were cooked. Because of its low digestibility, abraded peel would make little contribution to the nutrient requirements of high yielding dairy cows or fast-growing beef cattle. It is best suited to lower density diets such as those for young stock or suckler cows. This co-product could also be used in changeover rations for lambs which are being brought in from grazing and adapted to an intensive fattening diet.

At 10 per cent dry matter, 25 kg abraded peel could replace 2 kg average hay. It may be possible to feed larger amounts, and a reasonable target would be to replace up to one-third of the forage part of the diet. At such levels the abraded peel is likely to offer a palatable, succulent alternative feed, and it may be a useful supplement for animals kept at a relatively low plane of nutrition on fairly coarse roughages. Its protein content is significantly higher than that found in either hay or straw or even cereals, and this may be a useful feature when the co-product is used in conjunction with low quality roughages.

N.B. In contrast to all other potato co-products, abraded peel is low in starch and poses a low risk of acidosis when introduced to the ruminant diet.

Peel and trim

'Peel and trim' is a combination of co-products from crisp manufacture, separated in the early part of processing and marketed in mixed form. The material comprises a variable mixture of abraded peel and off-cuts of peeled, uncooked potato. Significantly, the mix may then be pressed and this removes a substantial proportion of free water, leaving a material with a much higher dry matter content and, consequently, a much greater value. A small proportion of the starchy sediment from the process water stream may also be added with the aim of producing a feed material with a consistent appearance.

For the manufacturer, this mixture represents a convenient means of disposing of both materials in a single operation, but the combination also offers advantages to the livestock producer. The abraded peel on its own has little place in the diets of highly productive stock, but the admixture with potato offcuts provides a considerable boost to the feed's energy value and makes the combined co-product a viable alternative feed. The peel moderates to some extent the high starch content of the potato trim and adds fibre, which may be expected to slow the overall rate of digestion and reduce the risk of acidosis. However, because of the variable proportions of the two components, it would be prudent to ration 'Peel and trim' in the same way as potato hopper. If a proportion of process water fines (feed grade starch) is added, this would increase the starch content of the blend and reinforce the need for prudence when feeding. Even without the starchy supplement, 'Peel and trim' represents a major upgrading of the nutritional value of abraded peel – a comparison of the starch content of a limited number of samples reveals a four-fold increase.

'Peel and trim' is best suited to the ruminant market. It is a solid material that can be stored in a bunker silo and used as a direct replacement for cereal grain. At 32-36 per cent dry matter, 2.75 kg 'Peel and trim' could replace 1 kg rolled barley. 'Peel and trim' needs to be introduced into the ration gradually and an adequate balance of fibrous feeds would need to be maintained.

Primary sludge (uncooked) / feed-grade starch

Primary sludge from the crisp manufacturing industry is a starchy residue similar in gross composition to the potato puree produced in chip manufacturing plants. The principal difference is that this sludge, which is separated from the cold process water as it leaves the factory, has not been subjected to any heat treatment. Larger pieces of potato are screened off and the fine co-product that remains largely represents the insoluble contents of the cells, which are released when the potato is sliced into thin sections. At low temperature, the starch grains have little tendency to absorb and chemically bind with the surrounding water, and much of that water can be separated from the primary sludge by centrifuging. The typical dry matter content of the centrifuged co-product sold as animal feed is close to 50 per cent and, in this form, it bears little resemblance to the sludge implied by the name. It would be more appropriate to regard it as feed-grade starch and this would certainly appear to be a more attractive feed name.

This co-product is virtually uncooked, which arguably makes it a feed more suited to ruminant rather than to pig feeding. However, after being subjected to large volumes of water, the water-soluble anti-nutritional factor present in raw potato is likely to have been substantially diminished in concentration, if not inactivated as it may have been in hot water. This remains to be demonstrated, but anecdotal evidence confirms that the material has been fed to pigs with satisfactory results.

In the ruminant animal, uncooked potato starch is digested more slowly than equivalent cooked products, but more quickly than barley, wheat or maize grain (Beever, 1999). Thus care is needed when primary sludge is fed to ruminants, and its incorporation into a mixed feed with an adequate proportion of fibrous supplements is recommended; 5 kg of primary sludge could replace 3 kg of rolled barley.

Potato crisps

Potato crisps are a highly digestible, oil-rich feed that can make a valuable contribution to the diet of cattle, sheep and pigs. The crisps leave the friers at a temperature of around 100°C and at that point the co-product is essentially sterile. It is also fully-cooked, with the potato starch fully gelatinised and the heat-labile protease inhibitor (present in raw potato) inactivated.

The high oil content of crips combined with a high starch level makes this an unique feed product – only oil and feed-grade fats can offer richer energy sources, and those materials cannot be offered without incorporation into other feeds. Potato crisps are a palatable, energy-rich feed that can be offered to ruminant animals either as a single feed or in combination with complementary feeds. For pigs, the high sodium content of crisps makes it advisable to feed this co-product as part of a balanced diet. Although the protein content of crisps is relatively low, protein quality is high – potato protein has more than twice the lysine content of cereal grain protein.

The high energy content of potato crisps makes the product suitable for high performance animals that need a diet with a high nutrient density. Crisps could be of particular benefit to high yielding dairy cows whose energy needs in early lactation outstrip their appetite. High ambient temperatures can also create a niche market for oil-rich feeds; under such conditions the animals' normal response is to restrict feed intake, which relieves the heat burden by reducing metabolism within the tissues, but this is achieved at the expense of animal performance. The replacement of some of the dietary carbohydrate by oil has a double benefit, increasing the energy intake some two and a half fold, and reducing the heat burden on the tissues as a consequence of the greater efficiency of fat synthesis from pre-formed fatty acids. Chudy and Schiemann (1969) have calculated that the replacement of carbohydrate by oil reduces heat production in the tissues by 63 per cent per MJ.

In common with other oil-rich feeds, potato crisps must always be rationed and fed with care to ruminant animals in order to avoid disturbance of microbial fermentation in the rumen. When formulating a ration containing crisps, the oil content of all other feeds must be taken into account, in order to control the oil content of the whole diet. Selection of the optimum dietary oil supply will require a consideration of the proportion of polyunsaturated fatty acids, since these have a greater effect on rumen microbes than saturated fatty acids. Dairy nutrition advisers often suggest a restriction of the oil content to a maximum of five per cent of the dietary dry matter, but there is a volume of evidence in the scientific literature that indicates positive responses in milk yield and milk fat content with diets containing up to eight per cent oil (Wu and Huber,1994). However, it is important to note that milk protein levels tend to decline in response to increasing oil intake before milk yield or fat level are affected.

For pigs, the high oil content of potato crisps offers specific benefits to both lactating sows and those in late pregnancy. The provision of dietary oil before farrowing has been shown to improve piglet survival. Oil used in the diet of lactating sows has increased milk supply and this has led to a greater weight of weaned pigs, when compared with an equivalent amount of energy supplied as carbohydrate

(Moser *et al.*, 1978; Pettigrew, 1981; Seerley, 1984). As with the dairy cow, the benefits of oil are seen more starkly under conditions of heat stress, but with pigs this benefit can have wider application because the temperature in the pig house is often deliberately held at a relatively high level for the benefit of the suckling piglets.

Dehydrated potato production

Dehydration is probably the oldest method of food preservation since it has long been recognised that the removal of water from food reduces the risk of microbial spoilage. Drying also brings other commercial advantages, including reductions in weight and volume with consequently smaller packaging and storage requirements and lower transport costs. There may be other effects too; drying may affect colour, texture, flavour and nutritional value. Not all of these consequences are of significance to the animal feed market, but a palatable, nutritious co-product with a relatively low haulage cost would have considerable attraction.

Potato dehydration began in Germany in the late 1800's and continues today in the form of flakes, granules, slices, diced potato and flour. Flakes are widely used in the snack food and ready-meals markets, whilst granules are also used for snacks and for reconstitution as potato mash. Sliced and diced potato and potato flour are all used by the food industry in the preparation of soups and ready meals, and some potato flour is also included in snack foods.

In the early days, dried potato was normally stored in this form and reconstituted before use. It was a staple ingredient of institutional diets such as those used in prisons and hospitals, and a limited number of niche markets were identified such as the use of deyhdrated potato as a food for campers. In the 1970's, the demand for dehydrated potato was given a considerable fillip by the development of the "artificial chip" – potato shapes formed from reconstituted potato. The market leader in this part of the snack-food industry was the "Hula Hoop", strips of reconstituted potato formed into rings and, more than twenty years later, came the "Pringle", a regularly formed and sized crisp made from reconstituted potato. The continued popularity of these products has established a secure outlet for the dehydration industry.

Co-products of dehydrated potato production

Potato flakes, granules, slices, and flour

Dehydrated potato is available in a variety of forms, all of which may be considered by the animal nutritionist to be equivalent to dry forms of potato mash. They are highly digestible cooked feeds comprising more than 75 per cent starch, but also containing a significant amount of good quality protein, a little fibre and some minerals.

Dehydrated potato can be used to replace cereals in rations for both pigs and cattle. For non-ruminants the superior lysine content of potato protein adds to the attraction of the feed, and its digestibility is enhanced by the inactivation of the anti-nutritional factor that is present in raw potato. For cattle the potentially rapid digestibility of the starch needs to be anticipated and excessive consumption avoided in order to avoid the danger of acidosis. In common with potato mash, caution needs to be exercised in the provision of a diet for ruminants that contains adequate amounts of long fibre distributed at intervals through the day. Matching the starch with rapidly available nitrogen sources may be expected to improve micriobial protein production in the rumen.

Up to 5 kg dehydrated potato per head per day can be fed to dairy cows, preferably mixed into a complete ration that will be available throughout the day. Dairy youngstock are typically fed only modest amounts of non-forage feeds but care will need to be taken if dehydrated potato is to be included in rations that will be fed only once a day. In this situation, it would be advisable to restrict consumption to no more than 1 kg dehydrated potato per head per day, fed during the day after a roughage meal. Intensively-fed beef cattle could make good use of this type of feed; rations containing up to 40 per cent dehydrated potato may be fed once the cattle have adjusted to the diet – providing the feed is continuously available and balanced with an adequate proportion of palatable roughage feed. Care would also be needed if dehydrated potato were to be fed to sheep. Because of individual variation in feed consumption, it would be advisable to restrict the potato component of the diet to no more than 10 per cent.

Potato canning

Canning was invented almost two hundred years ago when Nicholas Appert developed the system to preserve food for Napoleon's troops, and it was soon adopted by the British Admiralty after a link between diet and the onset, or avoidance, of scurvy had been demonstrated. The technology was not without its problems, however, and in the mid-1800's a large quantity of canned naval stock was found to be putrid. Throughout the first hundred years and beyond, reservations about the canning process were widely debated, and it was not until the 1920's that food canning became established in the UK (Hutton, 2001).

In a parallel to the assurance statements that are now sought for UK feeds, the Food Committee of the New Health Society provided the following testimonial for canned foods in 1929...

> 'Experience proves that modern, meticulous methods of harvesting, selection, preparation and canning guarantee absolute safety.'

The animal feed industry rarely experiences or expects "absolute safety", and it should quickly be noted that such a guarantee could not be provided for the co-products of canning. But it would be fair to claim that much of the same meticulousness that is ascribed to the process up to the canning stage applies to the fractions used for feed as much as to those entering the cans for human consumption.

Potato canning involves a minimum of processing; washed potatoes are steam-peeled before being washed again, and then packed into cans and cooked. The potatoes are not chopped or sliced and consequently no off-cuts become available as they do in other processing operations. The cooked product is completely secured inside the can and there is no possibility of a cooked co-product appearing at this stage. Thus the animal feed interest is concentrated on the early stages, and two key features of potato canning have a bearing on co-product output.

Firstly the potatoes used for this process are new and very small, and this implies a large surface area per unit weight - although new potatoes have a relatively thin skin, the "peel loss" from a tonne of small potatoes is relatively high. This fraction is also boosted by the use of a steam peeling procedure that removes more of the inner potato flesh than would be separated by simple abrasion. The choice of the peeling process is dictated by the requirements of the final product, which would be considered unacceptable if it still bore fragments of skin, as might be the case with abrasive peeling. In the author's experience, this requirement has also led to deeper peeling - so that the **potato feed** co-product contains a higher percentage of starch, and consequently represents a more digestible, higher energy feed material.

Secondly, the canners' objective is a perfect, blemish-free product. So, having removed every trace of potato skin, he then rejects a relatively high proportion of peeled potatoes because of minor imperfections. In other potato processing sectors, the eyes and other significant blotches would typically be cut away, leaving the rest of the tuber for further processing. This option is not available to the canner, since his specification is for whole tubers not potato pieces, and the small size of his raw material would make this difficult anyway. Thus, as a co-product of the canning operation, the potato processor produces a relatively large proportion of **peeled canning potatoes.** The minor blemishes that they bear are of no significance in animal feeding, and canning potatoes represent a clean and potentially trouble-free co-product. Their composition is similar to the off-cuts produced in other factories, though they contain virtually no potato skin, and their size makes them much less likely to cause the choking problems sometimes experienced with larger tubers.

Potentially, there is additional co-product material that arises when the newly-peeled potatoes are washed to remove the sticky outer coating. This predominantly starchy material disappears into the processing water, and it could be recovered in similar fashion to the puree that is produced in the potato chip and to a lesser extent in the crisp manufacturing industries. However, the lack of cutting in the potato canning industry, means that the "feed-grade starch" is available only in small quantity, and its recovery at present is not economic.

10 Sugar Beet Processing

It is believed that sugar was first harvested for its flavour some five thousand years ago when it grew as sugar cane on a number of Pacific Islands, but sugar did not reach Europe for another four thousand years until the time of the crusades (Lewicki, 1997). The extraction of sugar from beet is much more recent, being first achieved in 1748 following research by German scientists, and it was in Germany that the first sugar beet factory was opened in 1799. The development of the British sugar beet industry was later still, with the first factory being opened only in 1912 (British Sugar, 1998). Over the next twenty years the industry expanded across England, and by the mid 1930's there were thirteen companies processing sugar beet at eighteen sites. In 1936 an Act of Parliament (the Sugar Industry (Reorganisation) Act) amalgamated all of these companies into just one, the British Sugar Corporation (now British Sugar), and this organisation has been responsible since that time for managing the entire domestic crop.

Records of the amount of sugar beet processed by British Sugar are available from that time, and the information is presented in summary in Table 1. Individual years are shown for the pre-war years, but the post-war data represent three-year averages. The record shows a continual increase in the amount of sugar beet processed since the low point in 1938. An attempt has been made to calculate the accompanying growth in animal feed production and these figures are also shown in the Table. The type and the proportions of different sugar beet co-products have varied through the years, but the data presented in Table 1 have all been expressed in terms of the equivalent weight of dried sugar beet pulp.

Field operations

The quality of the end products of sugar beet processing is ensured by a quality management system that encompasses the entire operation from harvested beet through to the packaged product, and this system benefits the co-product feeds as much as the sugar products. However, the safety of the end-products is assured

Table 1: Development of sugar beet processing in Britain

Year	*Sugar beet sliced million tonnes*	*Feed production* '000 tonnes*
1936	3.6	320
1938	2.3	203
1948	3.9	336
1958	5.6	496
1968	7.1	630
1978	7.8	689
1988	9.0	804
1998	10.6	945

*Figures expressed in terms of dried product
Source: John Smith, British Sugar (personal communication)

at an even earlier stage since all sugar beet growers are required, as part of their contracts, to comply with the guidelines set out in the "Green Book" (The UK Pesticide Guide, 1999). Confirmation of satisfactory farm practice is provided by a field audit in which some 5-7 per cent of fields are audited each year as part of British Sugar's ISO 9002 accreditation. Final confirmation is available from the results of pesticide residue checks on both products and co-products, which include analysis for the active ingredients of every chemical used in the field.

The provision of research and agronomy services has also helped to improve the crop in the field, and a continuous programme of development has benefited the environment as well as the grower, by reducing nitrogen wastage as well as enhancing food safety. The development of effective fungicide and pesticide seed dressings has enabled a substantial reduction to be made over the last six years in the number of field applications and the amount of chemicals used on farm.

Today, sugar beet is grown by 9,000 arable farmers in Britain and the crop is transported to seven factories, principally located in the eastern counties but also in the West Midlands. An annual volume of approximately 10 million tonnes of beet is processed in a six-month season between September and February. The principal objective is the annual production of 1.3-1.5 million tonnes of sugar, but the removal of the sugar leaves about four million tonnes of extracted pulp and a further 0.3 million tonnes of unrefined extract. The extracted residue has a very high water content and further processing is needed before the pulp becomes available for use as an animal feed – much of it being dried and pelleted before sale. British Sugar produces around 900,000 tonnes of dried sugar beet feed, mostly molassed, and 120,000 tonnes of moist pressed pulp. The industry produces some 300,000 tonnes of molasses, although much of this is re-combined with the pulp and dried. Approximately 50,000 tonnes of molasses are available separately.

Up until the 1980's, the sugar beet crop yielded a substantial volume of additional feed in the form of the sugar beet tops (crown, stem and leaves). On those beet-growing farms that also carried livestock, strenuous efforts were made to improve the utilisation of a material that was prone to soil contamination. The incentive was immense; an average of 40 tonnes of tops per hectare meant 800 tonnes of feed for the average grower with 20 hectares of crop. Although the nutritional value of the tops was relatively modest – with a typical dry matter content of 16 per cent and an estimated ME value of 10 MJ/kg dry matter – the entire crop, if harvestable, was in excess of 6 million tonnes. Some of this crop was never harvested but ploughed back into the ground, and the proportion that was fed gradually declined as mixed farms became replaced by specialist arable units. During the 1980's, the two-stage harvesting of sugar beet, firstly the tops followed by a separate operation to lift the roots, was replaced by a single-pass system. The new system offered clear advantages for the harvesting of the more valuable roots, but the inevitable consequence was that the tops were all left in the field. In some areas, the tops continue to be grazed by sheep but the utilised proportion is now thought to be less than five per cent (S. Todd, British Sugar; personal communication), with the other ninetyfive per cent ploughed in as a source of organic matter and potash. Thus a potentially valuable feed material has largely been abandoned by the march of time.

Factory processing

In today's world, the sugar beet roots are harvested between September and December and metered into the factory according to a strict programme. Temporary on-farm storage of a proportion of the crop extends the processing season and allows the factories to operate as efficiently as a seasonal operation can be. On arrival at the factory, the roots are tested for sugar content and the 'tops and soil' tare is checked. Unacceptably high levels of these contaminants, or frosted and diseased beet, result in a rejection of the entire load and its return to the farm of origin. Acceptable loads are tipped onto dry ground at the reception site, or flushed out of the delivery vehicle by high pressure hose, and from there the beet is carried into the factory by means of water flumes. During this water-borne transfer any adhering soil becomes dislodged, and cleaning is encouraged by the rolling and tumbling of the beet as they rub against one another. In the flumes the sugar beet are floated away from any stones picked up by the harvester, and beet tops, weeds and other trash are removed by an automatic raking system which operates against the flow of the water.

After cleaning, the beet are carried up a large mesh (5mm) vibrating screen that removes dirty water together with the beet tails that have been rubbed off during transport. A smaller (2mm) screen then recovers the small pieces of beet from the water. At one factory (York), these beet tails – comprising 1-2 per cent of the weight of the incoming beet - are marketed to selected farms. (Beet tails are not available to sugar beet growers, because of the possible risk of the transfer of soil-borne plant diseases.) At all the other factories, the beet tails are processed on-site and their content of sugar and fibre becomes part of the main output. The soil is collected from the dirty water and disposed of in landfill or land reclamation projects. As a disease precaution, none is allowed to return to arable farms.

The cleaned beet are forced through a slicing machine, which operates in a similar manner to a kitchen grater. In contrast with the usual practice in potato factories, sugar beet are not peeled prior to further processing and the whole of the root is presented for sugar extraction. The slicer reduces the size of the roots from elongated globes weighing some 2-3 kg each to thin V-shaped slivers approximately 2mm thick. These slivers, known as cossettes, have a large surface area and are considered ideal for sugar extraction. The cossettes are mixed with water and held at a temperature of 70°C and a controlled pH for approximately ninety minutes. Heat treatment effectively sterilises the product and partially denatures the cell walls, which facilitates the release of sugar. The cossettes are moved slowly against a countercurrent of water that maintains a concentration gradient between the sugar beet cells and the surrounding water, and this encourages a continuous diffusion of sugar. The dark-coloured raw juice is separated and the extracted pulp, which at this point has a dry matter content of only ten per cent, is screw-pressed to remove some of the residual sugar and to increase the dry matter content of the pulp.

Much of the pulp will be dried on site and this is a crucial requirement given the vast scale of the operation - involving some 400,000 tonnes of beet each week during the season. Drying converts a perishable material into an attractive animal feed that can be marketed flexibly, in various forms, throughout the whole of the year and over a wide geographical area. Drying also offers other advantages, principal of which is the provision of a convenient and added-value route for the utilisation of molasses (see later). However, there is continuing demand for the fresh co-product, and potential economic attractions for the manufacturer if the material can be marketed without the cost of drying. The availability of a fresh market also acts as a safety valve for the factories in case of breakdowns. In consequence, it seems probable that a significant proportion of the extracted sugar beet will continue to be available in the fresh form – it is currently marketed as **pressed pulp**.

The raw juice is subjected to a purification stage known as carbonatation. Calcium hydroxide is added to the juice and carbon dioxide gas is injected into the mixture. The two chemicals react with one another and calcium carbonate precipitates from the solution, taking most of the impurities with it. This lime, with its significant trace element content, is marketed as a soil improver. The "thin juice" is concentrated by evaporation to about 65 per cent solids and the resulting "thick juice" is then stored for future use or filtered and transferred to the final crystallisation stage.

Crystallisation of the sugar takes place when the thick juice is boiled under vacuum and then seeded with tiny sugar crystals, which stimulate the formation and growth of other crystals. The resulting mixture of crystals and syrup is centrifuged and the separated liquor is then boiled again to produce what is curiously known as "raw sugar". The process is repeated a third time to yield a mixture of "final product sugar" and molasses. The raw and final product sugars are recycled into the thick juice, but the dark, concentrated molasses is marketed mainly as animal feed. Much of this molasses is mixed – contributing around 20 per cent of the dry matter - with pressed pulp and then dried to yield the popular and widely-known **molassed sugar beet pulp**. At one time, a proportion of this mixture was marketed in the undried form as molassed pressed pulp, but this moist blend is no longer available. A proportion of the molasses is marketed in its native form, as a heavily viscous syrup, usually referred to as **beet molasses** or simply **molasses**. Some molasses is marketed directly as an animal feed but much is used as a substrate in fermentation processes.

Industrial fermentation

A number of industrial processes employ molasses as a feedstock, making full use of the residual sugar content which comprises about 50 per cent of the syrup on a fresh weight basis (Karalazos and Swan, 1976). In essence, the sugar is used as the carbon and energy source for the growth of specific micro-organisms. Molasses has been the principal raw material in the production of baker's yeast - Saccharomyces cerevisiae - since the 1920's. Beet molasses is typically used in combination with cane molasses (10-20 per cent), and both contribute essential micro-nutrients as well as the major supply of carbon. Additional nitrogen in the form of ammonia, ammonium salts and urea is also supplied together with phosphorus in inorganic form. Other supplements may be added in specific situations, but molasses has been described as "not far from the ideal substrate for baker's yeast production" (Beudeker *et al.*, 1990). Yeast grown in this way is supplied to bakeries for bread making, and some is further processed to produce yeast extract, which is then spray-dried and marketed for use as a flavouring ingredient for the food industry. Since the development of molasses as an industrial feedstock for baker's yeast, a number of other fermentation applications have followed. These include the production of alcohol, single cell proteins, organic acids and even vitamins such as riboflavin.

After fermentation the residue of the syrup, together with yeast cell debris, is evaporated and marketed under the name **condensed molasses solubles** or **condensed molasses solids (CMS)** to the animal feed industry – though neither name appears to offer an adequate description of the material. Without the sugar, CMS is a much less viscous liquid, and some of it is re-combined with molasses to produce a blend that is easier to pump and to handle on farms. The removal of the sugar from molasses, together with the inclusion of the yeast fraction, boosts the protein content of CMS to the extent that a feed with only a modest protein content is transformed into a protein-rich supplement; c.f. 13 v 36 per cent crude protein in the dry matter. Less desirably, there is an associated concentration of the ash content of the feed, with notable increases in the sodium and potassium contents. This has a negative effect on the energy value of the feed, and the potassium content in particular may be potentially toxic if the feed were to be used at a high level (Karalazos and Swan, 1976).

The co-products of sugar beet processing

Molassed sugar beet feed

Molassed sugar beet feed (MSBF) is a co-product of sugar beet processing, and comprises a dried blend of two components: the extracted sugar beet pulp and the beet molasses that remains after sugar crystallisation. MSBF is the major co-product of sugar beet processing and is produced in very large volumes during a season, which lasts from September to February. Such is its popularity with livestock farmers that, on occasion, the supply has been limited for those wishing to use MSBF during the summer. This is a considerable change from the position that existed in the early days of British sugar beet processing, when the relatively unknown pulp was treated with some suspicion and much of the dried co-product had to be exported to find a ready market (Woodman and Calton, 1928).

MSBF is a high energy feed for ruminants, with the energy supply largely provided in the form of sugars and very digestible fibre. The feed's degradability in the rumen is slower and more continuous over a longer period than alternative energy-rich feeds such as rolled barley (ADAS, 1989), and its digestion typically produces a much lower amount of lactic acid. It may thus be considered a safer feed for high performance animals such as dairy cows that are being fed a highly concentrated, low-roughage diet. Sheep farmers rely on MSBF too, as a safer feed for ewes that are commonly fed in large groups, where individual consumption is difficult to control.

The fibre content of MSBF is highly absorbent and horse owners are repeatedly advised that this popular feed needs to be soaked for some hours prior to feeding – shreds for 12 hours and pellets for 24 hours (Harland, 1985). Horses are known to have a relatively small stomach, with an entrance that is controlled by a powerful valve that prevents both belching and the regurgitation of feed (Cuddeford, 1996). Soaking the pulp avoids the risk of problems caused by a potentially painful distention of the gut, which may occur when the feed is eaten in dry form but later swells as it absorbs water. However, in recent times, two new forms of sugar beet pulp have become available as a result of further processing and they do not require prolonged soaking. "Ultra Beet" is an extruded pulp that has been expanded during manufacture and needs no soaking (BSF), whereas "Speedi-Beet" is a micronised feed that can absorb water in less than ten minutes – much more quickly than the standard co-product (MMF).

The absorptive characteristic of the feed is put to good use in silage making when the addition of approximately 5 per cent of MSBF dramatically reduces the flow of nutrient-rich effluent from ensiled wet grass (Offer and Al-Rwidah, 1989), and enhances the feeding value of the conserved product (Crawshaw, 1991). A similar use has been developed for the co-storage of MSBF with brewers' grains or draff – an on-farm mixture that is known as Grainbeet. Here, the MSBF comprises about 15 per cent of the mixture, and the benefits have been measured not just in terms of juice retention but in reduced in-silo losses and improved levels of animal performance also. A research study by Hyslop (1991) found the in-silo loss of dry matter was reduced from 13 to 6 per cent when MSBF was added to draff prior to storage, and this saving reflected more than simply a retention of the juice. When draff and Grainbeet were fed to young Friesian steers, the Grainbeet diet stimulated a 13 per cent higher intake and an enhancement of growth rate from 1.05 to 1.2 kg per day. Interestingly, the performance of the Grainbeet-fed cattle was marginally better than that of a further group in which the animals were offered the two feeds separately.

Dried sugar beet pulp is a palatable feed for pigs, sheep and cows
(courtesy: Roquette UK)

Beet pulp in silage
(courtesy: British Sugar plc)

Many dairy herds are given a mid-day feed of 2-3 kg MSBF per cow, but much more can be fed as a direct replacement for rolled cereals. Castle (1972) fed concentrate rations containing 0-80 per cent MSBF, including up to 5 kg MSBF per cow, with incremental amounts replacing similar amounts of barley. No differences were recorded in either milk yield or quality, and the author concluded that the two feeds were of equal energy value. Van Es and colleagues (1971) carried out energy balance studies to determine the net energy of unmolassed sugar beet pulp, and their results confirmed that the co-product's digestible fibre content was utilised as efficiently as the digestion products of starch. However, the gross energy value of MSBF (17.1 MJ per kg DM) is relatively low and, although highly digestible (84-89 per cent), the commonly accepted ME value of 12.5 MJ per kg DM seems reasonable (figures from MAFF, 1990). Since the molasses content has a gross energy value of only 15.3 MJ per kg DM, it is axiomatic that the GE of molassed beet pulp would be lower than that of the unmolassed co-product. It may also be expected that unmolassed sugar beet pulp would have a higher ME value, although this would depend on the actual ME value of molasses, and this has proved difficult to determine – see discussion by Givens and colleagues (1992) in relation to cane molasses. MAFF Tables appear to confirm the higher energy value of the unmolassed co-product - suggesting an ME value of 12.9 MJ per kg DM. Such valuations place sugar beet pulp, in both forms, a little lower than barley as an energy source; barley has an average ME value of 13.3 MJ per kg DM.

Castle (1972) also reported a four-month study in which MSBF constituted approximately 50 per cent of the dietary dry matter for four cows in mid-lactation, and the maximum daily intake of MSBF was measured at 9.25 kg per day. Satisfactory performance was recorded for these cows, in terms of milk yield, milk quality and the maintenance of liveweight. Significantly, no reference was made to any problem of taint - an issue that had exercised the minds of early researchers (Cranfield and Mackintosh, 1935). Traditionally, it was the usual practice to soak dried sugar beet pulp before feeding it to dairy cows; a practice that continues to be recommended for horses. However, the consumption of substantial amounts of unsoaked sugar beet pulp in Castle's studies demonstrated that soaking the pulp was unnecessary for dairy cows. An increased demand for water was noted when rations including dried beet pulp were fed but, providing free access to water was offered, the cows were able to adjust the amount they drank without any detriment to milk production.

Beef cattle have fattened successfully on diets containing a high proportion of MSBF. Conclusive proof of this can be found in the study carried out by Cahill and colleagues (1966), in which MSBF constituted the bulk of the diet almost from birth to death. From 1-12 weeks of age an ad libitum diet was offered containing 76.8 per cent MSBF, from 12-24 weeks the MSBF comprised 84.8 per cent of the diet and, in the final phase from 24 weeks to slaughter at 380 kg liveweight, the inclusion rate was increased to 89.8 per cent. The Friesian and Friesian cross Hereford steers in this trial grew at an average of 1.05 kg per day, only marginally slower than similar cattle fed either barley (1.11 kg per day) or maize (1.08 kg per day) in place of the beet pulp. In a much larger study involving 702 cattle, Boucqué et al (1976) compared finishing diets containing 50, 60 and 70 per cent unmolassed dried sugar beet pulp for bulls on either an intensive or a semi-intensive system. The growth rates were similar in all cases and, at 1.17-1.36 kg per day, were considered to be satisfactory.

It is recognised that sugar beet pulp is deficient in protein if it is to be used at a high level in the diet of beef or dairy cattle. A number of trials with beef animals have shown that this deficiency can be met by the incorporation of urea as an economical source of supplementary nitrogen (Randall *et al.*, 1972, Hemingway *et al.*, 1976). For dairy cows, and particularly for those fed grass silage as a basal ration, vegetable protein supplements may be preferable to those based on urea. However, Parkins and colleagues (1974) quoted a number of reports of satisfactory milk production with the use of non-protein nitrogen supplements to a sugar beet pulp-based diet. Indeed, for a time, British Sugar marketed a form of MSBF – known as Triple Nuts – in which urea was included at 2.8 per cent.

The molasses content of MSBF may stimulate an improvement in nitrogen utilisation compared with the unmolassed pulp, at least when fed in conjunction with a rapidly available nitrogen source such as grass silage (Givens *et al.*, 1992). This effect was demonstrated in a metabolism study with Friesian cows by Beever and colleagues (1988), who compared both molassed and unmolassed sugar beet pulp with barley. They estimated the absorbed protein supply to be 11.3, 10.9 and 9.5 g per MJ ME for MSBF, barley and unmolassed SBF respectively. In a linked study (Sutton *et al.*, 1988), it was found that the yield of milk solids tended to be higher with both forms of beet pulp than with barley, but only in weeks 4-8 of lactation and only at the higher of two protein intakes. The authors concluded that the effects of carbohydrate source are greatest in early lactation, and that adequate protein supplementation of sugar beet feed was important.

Mineral supplementation should also be considered; whilst sugar beet pulp is a useful source of calcium it has only a poor phosphorus content. If it is used as a replacement for rolled barley, each kg of MSBF will supply 2.8g less phosphorus. This deficiency can be remedied by the use of additional amounts of mineral phosphorus, and this was the rationale behind British Sugar's inclusion of 3 per cent dicalcium phosphate, plus trace elements and vitamins, in the now defunct Triple Nuts (Parkins *et al.*, 1974). However, in place of such direct supplementation of an ingredient, it may be more appropriate to review the mineral status of the whole diet. Where MSBF is fed in conjunction with brewers' grains, there is a synergy in terms of the mineral supply of the two feeds. This synergy is seen to good effect in Grainbeet, the on-farm blend in which these two feeds are stored on many farms. In contrast with calcium phosphorus ratios of 9.9:1 for MSBF and 0.7:1 for brewers' grains, Grainbeet has a much more balanced ratio of 1.3:1.

Sheep are less able to maintain optimal rumen conditions than cattle, and this can lead to poorer fibre digestion and wasteful protein utilisation. MSBF has been shown to provide greater stability in rumen conditions than a supplement of rolled cereals (Rymer, 1988), and the ewe responds by consuming larger amounts of forage. This helps the animal to meet the high nutrient requirements of late pregnancy and early lactation. Hemingway and Parkins (1972) fed molassed sugar beet pulp as the sole concentrate feed to ewes and Dickson and Laird (1976) used it to replace the entire hay ration for pregnant ewes. Satisfactory performance was achieved in both trials providing that an adequate protein allowance was supplied. Sugar beet pulp is also well established as a supplementary feed for fattening lambs (Fairbairn, 1974). Bhattacharya and colleagues (1975) compared diets containing up to 90 per cent sugar beet pulp for fattening lambs, and obtained significantly higher growth rates than a comparable diet comprising 90 per cent maize. All of these sheep-feeding trials demonstrate the successful use of dried sugar beet pulp under controlled conditions but often, in practice, this co-product feed is chosen for more pragmatic reasons. Many sheep farmers regard the large size of the pellet, together with its palatability, as the feed's most useful attributes. These characteristics allow large groups of heavily pregnant sheep to be fed relatively small amounts of supplementary feed, scattered on the pasture with reasonable assurance of an even distribution between animals.

MSBF has also been shown to be a useful feed for pigs. The bacteria that are predominantly responsible for fibre digestion in the rumen have been found in large numbers in the large intestine of the pig (Varel *et al.*, 1984), and the proportion of cellulose-digesting organisms increases with an increasing level of fibre in the diet. Non-starch polysaccharides are the most useful definition of fibre for non-ruminant animals, and this fraction comprises more than 50 per cent of the MSBF. It has been estimated that up to 30 per cent of the energy used by sows and growing pigs can be derived from the fermentation of non-starch polysaccharides (Longland *et al.*, 1990), which clearly allows considerable scope for the use of MSBF.

In farm-scale feeding trials, both the growth rate and feed conversion efficiency of growing pigs were optimal at an inclusion rate of 20 per cent MSBF, with a marginal decline at 25 per cent (Kay and Simmins, 1990). Sows may have a greater potential than growing pigs to utilise a fibrous feed, and this hypothesis was investigated by the same authors. They found that 40 per cent MSBF could be included in sow diets without any reduction in reproductive performance, but the sticky consistency of the dung presented cleaning problems in the pens. For this reason the authors suggested that an inclusion level of around 20 per cent may be preferable in practice. In a review of recent research work on the role of sugar beet pulp in the nutrition of pigs, Close (1995) concluded that 15-20 per cent of MSBF in a growing pig diet gave a comparable growth rate and feed conversion efficiency to cereal-based rations, but sugar beet pulp improved the eating characteristics of the meat. In sow diets, he found that 20 per cent MSBF during pregnancy and 10 per cent during lactation gave superior performance to cereal-based diets - in terms of litter size, milk yield and composition, piglet performance, speed of re-breeding and sow welfare.

Pressed pulp

Pressed sugar beet pulp is the fibrous residue that remains after the extraction of sugar from beet and a reduction of the initial water content. The short-hand title of pressed pulp is widely used and understood in the UK, but a number of other pulps are also used as animal feeds, such as those of cereals, citrus, manioc and potato. A little confusingly, the draft list of ingredients for compound feeds (EC, 1990) even contains "molasses pulp", though its description suggests that it would be known in the UK as dried molassed sugar beet pulp. Much of the pressed sugar beet pulp will be dried within the factory, usually after mixing with molasses, but some is marketed in the fresh form. The cost of drying represents a major financial burden on the factory, and the gaseous emissions may constitute a potential environmental risk. Thus the sale of pressed pulp, a co-product of slightly superior nutritional value to dried sugar beet pulp (ADAS, 1979), is a valuable alternative for a proportion of the production and some 4 per cent of the pressed output is currently marketed in this form. Improvements in the screw pressing of the pulp have resulted in a material with a higher dry matter content, and this has benefited both the drying operation and the value of the fresh co-product. Whereas the pressed material formerly had a dry matter content of around 18 per cent, the pressed pulp now being marketed averages 25 to 26 per cent, although there are variations between factories.

For the livestock farmer, pressed pulp represents an attractive, succulent feed of high nutritive value. Although this co-product has a high fibre content (with a crude fibre level of almost 20 per cent), that fibre is of high digestibility – an average of 85 per cent in studies at the ADAS Feed Evaluation Unit (MAFF, 1990). The ME value of 13.0 MJ per kg DM was determined in an *in vivo* study at the ADAS Feed Evaluation Unit (Trident Feeds FF3). At this level, the ME value is marginally higher than that of either the molassed or unmolassed dried co-products (12.5 and 12.9 MJ per kg DM respectively). Pressed pulp may be used in a wide range of ruminant diets, and may even play a role in pig feeding. Because its energy supply is in the form of digestible fibre, pressed pulp may represent a safer feed for ruminant stock than more starchy alternatives that can be associated with problems of acidosis. Pressed pulp can thus be particularly useful as a palatable ingredient in diets for high yielding dairy cows maintained on high concentrate rations.

In common with many fresh co-products, pressed pulp is prone to aerobic decay and, unless the feed is to be consumed within a few days, it needs to be ensiled to preserve its attractive qualities. Safely stored, the ensiled co-product may be regarded as of equal value to the fresh material. There has even been an isolated report (ADAS, 1979) in which the ensilage of pressed pulp was shown to increase its digestibility, but this unexpected finding has never been followed up. Recent comparisons of fresh and ensiled pressed pulp have suggested that the ensiled co-product may be more palatable to cattle and sheep (M Witt, Trident Feeds; personal communication).

Dairy cows are commonly fed up to 20 kg of pressed pulp in place of an equivalent amount of maize silage or rolled barley. In a trial at Glasgow University (Harland, 1981), 3 kg of dried sugar beet pulp were replaced by an equivalent amount of pressed pulp in a changeover trial. In both groups, the cows receiving pressed pulp gave a little more milk than those on the dried feed. Adequate levels of long forage are considered advisable in dairy cow diets, even in those containing large amounts of pressed pulp, in order to stimulate saliva flow and to reduce any risk of rumen acidosis. Attention should also be paid to the achievement of a good mineral balance – sugar beet pulp co-products are typically rich in calcium but low in phosphorus. In mineral content as well as energy and protein levels, pressed pulp is complementary to brewers' grains and the two co-products can be ensiled together. More commonly, however, livestock farmers ensile a blend of brewers' grains and dried sugar beet pulp, and there is experimental proof that this mixture, popularly known as Grainbeet, is nutritionally superior to the sum of its parts.

Pressed pulp has been offered ad libitum to Friesian steers in the final phase of an 18-month beef system, with various concentrate supplements fed at 3.5 kg per head per day. Satisfactory growth rates of up to 1.4 kg per day were achieved, and intakes of 31-36 kg of pressed pulp (of 17 per cent DM) were recorded (Norfolk Ag Stn, 1981). This amount would be equivalent to perhaps 20-24 kg of today's higher dry matter co-product – a dry matter intake of 1.2-1.4 per cent of liveweight. Pressed pulp has also been fed in place of maize silage to Charolais, Simmental and Sussex cross finishing cattle (Harland, 1981). The mean growth rate of the pressed pulp group, at 1.56 kg per day, exceeded that of the maize fed group, and animal health, killing-out percentage and carcase quality were all acceptable. These trials indicate that pressed pulp, in conjunction with a little straw, may replace high energy concentrate feeds and much of the dietary forage for fattening cattle where high rates of gain are required.

Pressed pulp is sometimes chosen by sheep farmers who value its safety and palatability as well as its high energy content. Most sheep are generally unaccustomed to concentrate feeding and even small quantities of starchy feeds offered to groups of animals may cause digestive upsets because of wide variations in the amount eaten by individuals. Because of its palatability and bulky nature, as well as its provision of energy in the form of digestible fibre, pressed pulp can be used with more confidence to introduce sheep to trough feeding. Two to three kilograms (per head per day) of pressed pulp would be a typical amount used for breeding or lactating ewes and for fattening lambs.

In 1998, a new farm blend was promoted with some success; pressed pulp mixed at a 4:1 ratio with distillers maize to yield a well balanced feed mix suitable for feeding to high performance beef and dairy animals. The mixture, which is known commercially as "Praize", is similar in concept to Grainbeet. However, where the latter is a moist protein feed (brewers' grains) boosted by the inclusion of a dry energy source (MSBF), Praize is a moist energy feed enhanced by the addition of a dry distillery co-product rich in both protein and energy. In the original trial work, where Limousin cross steers and bulls averaged a growth rate of 1.2 kg per day, the blend also included 10 per cent of bread, but this has not proved essential. The moist blend of pressed pulp and distillers maize has a theoretical ME value of 13.5-14 MJ per kg DM and a protein level of 19.5 per cent of the dry matter.

Sugar beet tails

Beet tails are the small end-pieces of sugar beet roots that become detached as the beet are tumbled and rolled in the water flumes that carry the raw material into the factory. Although the beet tails, at one to two per cent of the total weight, may seem to be only a minor fraction, the vast volume of the sugar beet crop (~10 million tonnes) ensures that the tails are a co-product of potential significance. However, it is now common practice at most factories to utilise the beet tails within the process and they are not widely available outside. Sugar beet tails continue to be available from the York factory and, during the season, 20,000 tonnes of this co-product become available as animal feed. However, unlike every other product and co-product of sugar beet processing, beet tails are not subjected to the heat treatment that ensures the destruction of any disease organisms. Consequently, sugar beet growers are prohibited from receiving this co-product in order to prevent any risk of the spread of soil-borne diseases.

Sugar beet tails are a relatively unprocessed fraction of the crop, having been broken from the root ball and subjected only to a degree of washing and draining. In comparison with whole beet, the tails are more fibrous, higher in moisture and may be expected to have a lower sugar content. Owing to the relatively high surface area of the beet tails, and their manner of separation from the harvested crop, this co-product does have a variable soil content. This directly impacts on the nutritional value; thus the calculated ME value of 11 MJ/kg dry matter should be related to an ash content of around 15 per cent.

This co-product feed represents a moist concentrate of high digestibility and potentially rapid degradability. Sugar beet tails are a highly palatable material that needs to be fed with care to ruminant animals, and always balanced with adequate supplies of hay or silage in order to avoid acidosis developing as a result of the rapid digestion of the substantial sugar fraction. Beet tails could also be fed to pigs in partial replacement of the cereal component of the diet.

Beet molasses

Beet molasses is a co-product of the sugar beet industry, comprising the residue that remains after the crystallisation of sugar from concentrated sugar juice. Molasses is a heavily viscous syrup, still rich in uncrystallised sugar but with a significant crude protein content too. It is the crude protein level that most clearly distinguishes this material from the molasses produced as a co-product of sugar cane extraction – beet molasses has three times the protein content of cane molasses. Practically all of the crude protein in beet molasses is digestible and, in the ruminant animal, this protein is available for microbial synthesis.

Beet molasses is a rich source of highly digestible sugars and is primarily used as an energy source. It is an ideal complement to other ruminant energy feeds such as cereal grain and maize silage, whose energy supply is degraded more slowly. It is also a valuable supplement for grass silage since its rapidly degradable sugar content matches the readily available crude protein, leading to improved protein utilisation in the rumen (Givens *et al.*, 1992). Molasses has also been widely used as an additive during silage making – providing additional fermentable sugar to ensure the satisfactory preservation of the crop. However, the efficient application of molasses to harvested grass was not easily achieved in practice and, although potentially effective, molasses has largely been replaced in recent years by inoculants, acids and enzymes which can be selected from a long list of silage additives.

Many mixed rations include a small proportion of molasses, added for a variety of reasons but principally for its palatability and the expected stimulation of feed intake. The palatability and attractive aroma of molasses make it a valuable component in many situations, particularly where it is desirable to mask the less desirable characteristics of other feeds. Its effect on intake has been quantified by Gillespie (1985), who quoted a dairy cow trial at the Animal and Grassland Research Institute in which 1kg of cane molasses stimulated silage intake by more than 8 per cent. Molasses is also highly sticky, and this characteristic helps it to bind minerals and other dusty components, which may otherwise drop out of a mix or blow away in the wind. In the compound feed industry molasses is used as a pellet binder.

Molasses is also the basis of liquid feeding systems - used in the field to provide small supplements to grazing animals. Such "ball and lick" systems have been employed on many sheep farms where the amount of "hand-feed" is usually small in relation to the animal's appetite, and trough feeding is considered to be an undesirably rough experience for heavily pregnant ewes. Molasses-based "licks" have also been formulated and pressed into solid blocks, and these have been used for both sheep and cattle, to provide a small supplement to grazing stock. One specialist use of both these forms of "lick" has been the provision of a palatable source of supplementary magnesium for animals at risk of hypomagnesaemia, particularly those grazing spring grass.

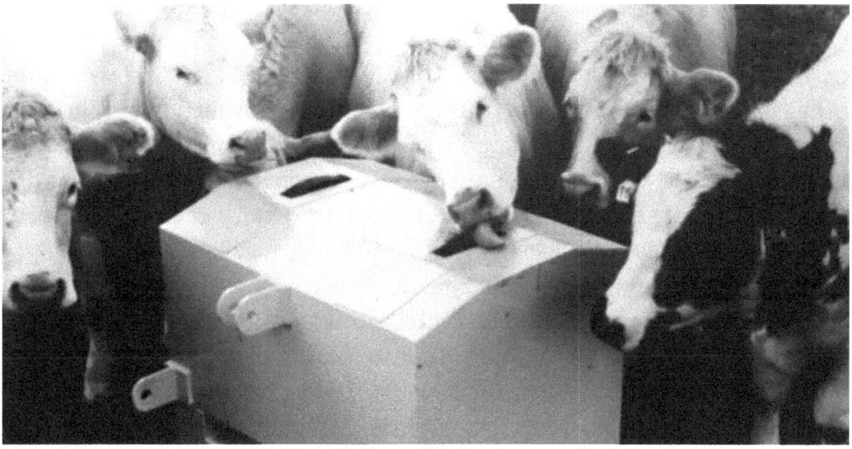

Molasses consumption can be controlled by a "lick-feeder"
(courtesy: UM Feeds Marketing)

The gross energy value of beet molasses is relatively low, at 15.3 MJ per kg DM, though that figure is in line with the gross energy value of various sugars. The ME value of beet molasses was determined in an in vivo study of six consignments at the ADAS Feed Evaluation Unit (ADAS, 1986). However, the molasses was fed at only 17 per cent of the dietary dry matter together with hay, and the authors commented that such a procedure inevitably results in an inherently high variability in measurement. Thus the mean ME value of 11.2 MJ per kg DM represented a range of values from 10.3 to 12.0.

Molasses has a relatively high ash content, and much of this is in the form of soluble minerals. This befits its origins as an aqueous extract of sugar beet from which much of the sugar has subsequently been removed. The high electrolyte content – including a particularly high potassium level - must be borne in mind if the feed is to be used for pigs at levels that may disturb their water balance (ARC, 1981). In some situations, a degree of concern may be appropriate with regard to the use of significant quantities of molasses for ruminants, for whom a high potassium intake may interfere with the absorption and utilisation of magnesium. Such an effect has been implicated in the aetiology of hypomagnesaemic tetany (Brouwer, 1952; MAFF, 1983). In contrast to the high level of soluble mineral elements, insoluble mineral elements are in poor supply; the calcium level is relatively low and phosphorus is almost absent. Adequate supplementation with these minerals would be needed for all classes of stock if molasses were to be used in significant amounts.

Molasses appears to be much less valuable as a feed when included at a relatively high dietary level. Research studies have measured substantially lower net energy values when the molasses inclusion rate for beef cattle was increased from 10 to 25 or 40 per cent of the diet, and from 10 to 30 per cent for dairy cows (Lofgreen and Otagaki, 1960). The principal cause of this dramatic reduction in value appears to be the consequence of a change in the end-products of digestion, towards a high butyrate and low propionate pattern in the rumen (Marty and Preston, 1970). This not only results in less efficient utilisation of dietary energy and protein, it may even induce a metabolic disorder with the animal being physiologically incapable of tolerating such a situation (Karalazos and Swan, 1976). High butyrate levels in the rumen result in a reduction in both glucose absorption and glucose synthesis, and an accumulation of blood ketones. In extreme, this situation leads on to ketosis and may result in death (Losada and Preston, 1974). However, ketosis is not an inevitable consequence and Preston and his colleagues have fattened beef cattle on diets containing 70-80 per cent molasses, though growth rates and feed conversion efficiency were relatively poor (see Karalazos and Swan, 1976).

Swan (1978) has suggested an upper limit of 20 per cent molasses in the ration dry matter, beyond which the rumen microflora may become dominated by sugar degrading types at the expense of fibre-digesting bacteria, with potentially damaging consequences. The optimum inclusion level may be somewhat lower. Karalazos and Swan (1976) reported unpublished work by Preston and his colleagues in Cuba, in which cane molasses was fed to fattening Zebu bulls in differing amounts. The fastest growth rate was achieved when molasses contributed only 11 per cent of the energy intake. In the light of these experiences, it would seem important to ensure that molasses is fed in restricted quantity as part of a well-balanced ration, when the feed can be expected to bring a number of significant benefits without any danger to health.

In the UK, dairy cows are typically fed 1-2 kg of beet molasses in mixed rations, but up to 3 kg per head have been used satisfactorily (Cox, 1978). In practice, lower yielding cows and rearing heifers may be fed about 1 kg molasses per head, often in conjunction with moderate quality roughages, whose palatability it could be expected to improve. Beef cattle may typically be fed 1-2 kg beet molasses, and at this level the feed could be expected to make a contribution towards the animal's energy needs, whilst enhancing palatability and total feed intake. Ewes in late pregnancy are sometimes fed molasses in the hope that the feed will boost blood sugar level and ward off the danger of pregnancy toxaemia (twin lamb disease). Since the molasses sugar is rapidly fermented in the rumen, it seems unlikely that the feed will have such a direct effect on blood sugars and, if molasses is fed in excessive quantity, it may have the opposite effect – for the reasons discussed above. However, the fact that farmers believe that molasses has a positive effect may indicate that appropriate amounts are being used, and there may be an indirect benefit exerted through an improvement in the ewe's energy intake - 0.1-0.2 kg beet molasses would be a typical allowance.

As well as playing a valuable role in ruminant diets, molasses can also be fed to pigs. It is typically included at about 5 per cent of the diet, principally as an energy feed that also reduces the dustiness of a dry mix and adds palatability to any ration. Much greater quantities have been fed; Walker (1985) recommended an inclusion of 15 per cent molasses in pig fattening rations as much as 37 per cent of the diet for in-pig sows. However, he commented that these levels were too high to be included in any meal and that the molasses would need to be given as a liquid feed. The greatest difficulty he encountered was a tendency for ad libitum fed pigs to consume too much molasses, and he suggested that a diluted molasses solution should be fed for only part of the day, with water being offered through the pipeline for the remainder of the day. For many pig farmers, the high levels of sodium and potassium in molasses would necessitate restricting the feed to a lower inclusion rate. Even where a separate supply of clean water was continuously available to enable the pig to excrete unwanted levels of electrolytes, there would be a strong likelihood of dirty pigs and an increased risk of disease transfer via the loose dung. The practical limit for molasses inclusion would be dependent on the use of other feed ingredients that may also include a high sodium content – a range of such feeds is available to pig farmers operating a liquid feeding system.

Condensed molasses solids

Condensed molasses solids (CMS) are a concentrated form of the liquor that remains after the fermentation of molasses in the production of yeast. During the fermentation, much of the sugar is used as an energy source for the growth of the yeast, and CMS comprises a sugar-depleted molasses residue that is considerably enhanced by the inclusion of yeast cell debris. The composition of this co-product appears to belie its name - CMS is a free-flowing liquid that is rich in protein and low in sugars, substantially different from the energy-rich, viscous material from which it is derived.

CMS is widely used in a blend with molasses, although it is a valuable feed in its own right. Blending serves to reduce the viscosity of the molasses, making it easier to use on farms, but CMS brings more than just a physical benefit to the blend; it substantially improves both the protein and the phosphorus supply. Compared to the original beet molasses, the crude protein level of CMS has been increased from 13 to 36 per cent, and the phosphorus content from virtually nothing to 0.7 per cent, on a dry matter basis.

The loss of the sugar during the fermentation of molasses results in an increase in the proportion of other components that remain with the residue. The relatively high ash content is of note and, in particular, the high electrolyte content - with 3 per cent sodium and 5.5 per cent potassium in the dry matter. These soluble mineral elements make CMS less desirable as a feed for pigs, which are sensitive to high sodium levels in particular. However, cattle and sheep can tolerate relatively high levels of sodium and potassium, and this co-product feed is more suitable for ruminant animals. It is a palatable feed with a high organic matter digestibility, and high levels of both phosphorus and magnesium too.

The high crude protein content has already been highlighted, though it should be noted that this fraction contains a high proportion of non-protein nitrogen. This reinforces the greater suitability of CMS for ruminant rather than non-ruminant stock. Karalazos and Swan (1976) reported a comparison of CMS with soyabean meal and with a molasses/urea mix. The soyabean diet promoted a greater flow of nitrogen through the duodenum and a higher nitrogen retention by the animal. It was concluded that CMS crude protein may be utilised with similar efficiency to that of a molasses/urea diet.

11 Wheat Fractionation

Cereal and potato starch have both been used in food preparation for hundreds of years, but in Western Europe starch was first employed I the textile industry as a stiffener for cloth (Kent-Jones and Amos, 1967). As recently as 1982, there was virtually no wheat fractionation industry in Britain, but there was an EU levy on the importation of hard wheat from the Unites States and this financial penalty stimulated interest in the extraction of gluten from European wheat, which could be used to supplement the flour used by British bakers. Less than 20 years later, an estimated 600,000 tonnes of wheat are fractionated each year by three companies located in London, the Midlands and Scotland. However, although gluten may have been the key ingredient that triggered this development, the starch fraction is now regarded as much more important. Quantitatively, this shift is emphasis is unarguable: wheat contains perhaps 8 per cent gluten but has a starch content of 60-75 per cent on a dry matter basis.

Once separated, the starch has a wide range of potential uses. There is demand from the food industry for its direct use as a food ingredient in a variety of products, including breakfast cereals and baby foods, pie fillings and confections, sauces and pickles. The paper industry uses it to enhance the strength and improve the surface properties of its products and the packaging and board industries choose it as an adhesive that is recyclable, safe and approved for food use. Some of the starch is converted to sugars and marketed in the form of syrups, and these are used for a range of purposes in baking, confectionery, pharmaceuticals, brewing ad soft drink manufacture.

Up to the initial extraction of starch, the fractionation of wheat is similar in all three British factories, but there are differences downstream. In London, the Greenwich factory saccharifies the remaining starch prior to alcoholic fermentation of the resulting sugars. In Scotland, a similar raw material is passed to the distillery that shares the same site, where it is saccharified and fermented as it is in Greenwich. At Corby, in the English Midlands, the third of these wheat processing companies does not convert the starch and ferment the sugar, but continues the separation of the remaining wheat fractions by centrifugation.

At the end of processing there is a residual material comprising a blend of sugars, acids, proteins, fibre and minerals. In Corby, this is a low protein material that is marketed as a feed energy source for liquid-fed pigs. At Greenwich, the wheat residues are enhanced by the yeasts that multiplied during the fermentation stage, and the residual co-product is marketed as protein-rich liquid feed for pigs. In Scotland, the wheat fractionation residues are combined with other residues of grain whisky and neutral spirit production. These contain a high proportion of yeast-enriched solids, and the feed co-product from this operation is a solid material that is obtained by a fine filtration process and it is marketed to the ruminant industry. Although these residual materials vary in composition, and cannot be further separated at economic cost, each one represents a consistent source of nutrients that is used and valued as a feed by livestock farmers.

Crop selection and protection

In terms of nutrient composition, the wheat fractionation industry is less exacting than the flour miller, whose requirements are more precise and his options more limited. To a much greater extent, the fractionation industry can utilise wheat with a wide range of composition and turn it into a collection of its component parts.

This crop will yield starch, glucose syrups and wheat gluten as well as animal feed
(courtesy: Roquette UK)

However, before input to the factory is permitted, the wheat must satisfy a number of checks to ensure its safety and suitability. Supplies are obtained only from assured farms (such as ACCS, or SQC in Scotland), at which growers comply with the requirements of the Pesticide Regulations (1998), Control of Pesticides (1986) and the Food Safety (1990) Acts, and compliance is checked and verified by independent inspectors. All consignments must carry a grain passport, which identifies the source and records the use of any post-harvest chemicals. Transport is by dedicated grain vehicles, which completely avoid any danger of cross-contamination with other commodities, or by outside hauliers who must comply with the UKASTA Code of Practice for Road Haulage in respect of cleaning routines and the exclusion of a proscribed list of goods.

On arrival at the factory, the grain is sampled and subjected to a number of tests before the vehicle is permitted to unload. Rapid laboratory procedures check the moisture, protein, bushel weight and admixture, and the wheat is organoleptically checked to ensure a suitable smell and appearance. Acceptable consignments are pre-cleaned to remove any extraneous materials before the grain is loaded into the silo. Any evidence of ergot would be detected at this stage and would lead to the rejection of the load. The grain is subsequently cleaned again to remove weed seeds and dust, and these waster materials are consigned to landfill. Mycotoxins and pesticide residues are tested as part of a routine monitoring programme, and a dossier of satisfactory results is claimed. Such records remain confidential to the companies that produce them, but an indication of the pesticide levels that may be expected on wheat grain can be seen in the test results provided by the millers' association, nabim, which are quoted in the Bread and Baking chapter. ". . . the great majority of samples examined contained no detectable residues, and only a tiny proportion (0.2 per cent over the last seven years) showed residues above the MRL's that reflect the limits of good agricultural practice".

Wheat processing

Although wheat processing begins as it does in the flour-mill with the grinding of the grain, roller milling is a gentler process which has a less damaging effect on the starch granules. The initial processing extracts the grain in a highly uniform manner, and this leads to a very consistent quality **wheatfeed** becoming available as a co-product for the animal feed market; (c.f. the flour mill where a number of different flour types is produced, with a consequent range in wheatfeed quality). The wheat germ is a very minor fraction, comprising less than one per cent of the wheat grain, but is may be separated at this point and used as a valuable ingredient in bread making. In other situations, the wheat germ is not marketed separately but added to the wheatfeed, and this is the case at the Greenwich factory.

Further processing differs to some extent between factories, but the basic fractionation process is as follows. Water is added to the flower to convert it into a thin, pancake-like batter and, at Corby, a relatively fibrous fraction is separated. In reality, this fraction is much richer in starch than fibre, but its removal facilitates further fractionation of the flour according to particle size. This flour fibre fraction, with a dry matter content of approximately 16-18 per cent, will later be recombined with another fraction and marketed as a liquid feed for pigs. The remaining batter (or the whole of it where the fibre is not separated) is then subjected to a series of centrifuging operations, which split the product into a number of distinguishable components.

Firstly the gluten is separated and, after washing and drying, this is marketed to the baking industry. Large grain starch (sometimes referred to as A starch) is also separated and after similar washing and drying it is put to a number of food and non-food uses. Within the wheat starch industry, undried A starch is also converted by enzyme attack into sugars and the resulting solution evaporated to yield a range of syrups. These are collectively referred to as “glucose syrups” although their composition does include other sugars. These syrups are mostly used in the food, brewing and fermentation industries, with a little used in pet foods.

After the removal of the A starch, the residual liquor contains a mixture of smaller starch granules (sometimes referred to as B starch), broken starch fragments, soluble wheat protein (i.e. non-gluten), some fibre, sugars, oil and minerals. At Greenwich, the centrifuging is also claimed to remove B starch and to leave even finer material, referred to as C starch. This residual material is then saccharified by enzymes and fermented by yeasts (Saccharomyces cerevisiae strains, similar to brewers yeast) after adjustment of the pH to the optimal range. At the Girvan wheat factory in Scotland, the whole of the B starch and other residual components are transferred to the neighbouring grain distillery, where subsequent processing takes place. The alcohol that results from fermentation is distilled and use din the production of potable spirits, such as gin and vodka – though whisky cannot legally be produced from this raw material.

At the London factory, the distillation residues contain virtually no starch, although a small amount is usually measured by the standard enzymic analysis. What these residues do contain is an appreciable amount of non-fermented and non-fermentable sugars, plus lactic acid, protein, oil and minerals from the wheat flour, in addition to a proportion of yeast cell debris. This liquor is partially evaporated, before being marketed as liquid feed – known as **Greenwich Gold™** - to the pig industry. In Scotland, the material remaining after distillation is passed through very fine filter presses that recover most of the suspended solids. This high protein material is marketed as a moist concentrate feed under the name **Vitagold™**. Since, in reality, this is a co-product of the distilled industry, Vitagold is described more fully in the Distilling chapter.

Where the factory operations include an adjustment of pH, in order to provide the optimum conditions for fermentation or enzyme activity, this commonly includes the use of either sodium hydroxide or sulphuric acid. Such additions may have significant implications for the pigs that are to be fed on the co-product, unless controlled adequately. On occasion, excessive sodium addition has been claimed to lead to sodium toxicity and pig deaths, particularly in situations where the pigs' access to drinking water had been interrupted. This potential risk is now well recognised, and the sodium content of the co-product is monitored closely, with careful adjustment to ensure the level is maintained within strict limits.

At the Roquette factory in the Midlands, the liquor remaining after the removal of the A starch is not fermented but further fractionated by centrifugation into three distinct components. The first of these fractions comprises the separable starch and a little soluble protein. After heating this material to expand the starch molecules, the pH is adjusted and enzymes are used to hydrolyse the starch into sugars. After evaporation, this co-product is marketed as a viscous syrup under the trade name **Roux™**. Its characteristic coffee colour shows some evidence of Maillard reactions having occurred between the sugar and the protein. However, this co-product does not have a high protein content and the animal feed trade typically uses Roux as a replacement for molasses. In comparison with this energy source, the much lighter colour of the Roux syrup is notable. The second faction to be separated by Roquette is essentially a concentration of the residual protein known as **Promanna™**, to be used in baby pig and other speciality compound feeds. The end-product of wheat fractionation in Corby is a brown-coloured liquid comprising the residual components of this process. It is partially evaporated before being mixed with the flour fibre fraction separated in the initial stages of processing, and this co-product blend is marketed as an energy-rich liquid feed for pigs under the name **Abracarb Plus™**.

Co-products of the wheat starch industry

Wheatfeed

Wheatfeed is a dry co-product, comprising a mixture of the bran and a proportion of the endosperm, separated when cleaned and dehusked grain is milled for flour. In traditional flour milling for the baking industry, wheatfeed quality varies with the type of flour being produced, but in the wheat starch industry the co-product is the consistent output of a consistent extraction system. Although there are differences between seasons, flour extraction is typically 76 per cent of the dehusked grain. Thus both the flour and the wheatfeed yielded by this process are slightly more fibrous than the equivalent materials produced by the miller when his objective is the production of white flour – normally achieved by a 72-74 per cent extraction of the grain.

Wheatfeed is as traditional as bread making and its qualities have long been recognised in animal feeding. It is widely used in both ruminant and pig rations where it is regarded as a useful source of both energy and protein. Wheatfeed is also a valuable source of phosphorus for ruminant animals but its availability to non-ruminants is typically low; viz: 18-35 per cent for pigs (Jongbloed et al., 1991). In addition to its nutritive value, wheatfeed is used in many pig rations to balance some of the more refined, low fibre products and, through its buffering effects, to help provide a stable gut environment that promotes a desirable hind-gut fermentation.

More recently, wheatfeed has become available in a moist blended form with Greenwich Gold and is marketed under the name Amyplus. This combination has widened the market for Greenwich Gold, and enabled it to be fed in a solid form to ruminant animals. Wheatfeed is traditionally used by a wide range of livestock farmers, and the principal advantage of its availability in the form of Amyplus would appear to be an enhancement of its nutritional value.

Greenwich Gold™

Greenwich Gold is a co-product of alcohol production in Amylum's wheat processing factory at Greenwich, and comprises the yeast-enriched material that remains after distillation. The material comprises a concentrated mixture of the residual parts of wheat flour after the extraction of starch and insoluble protein – the oil, fibre and ash contents are all increased compared with the original flour. But there are additional components that have been formed during the processing, such as lactic and acetic acids, and together with the yeasts these contribute significantly to the co-product's nutritive value. The liquor is partially evaporated in the factory to a relatively consistent dry matter content of 21-23 per cent, and it is marketed as a rich golden fluid that can be pumped and mixed easily. Greenwich Gold is more widely available than any other liquid feed used in the UK pig industry, and it has set the standard for other liquid feeds since 1992. More recently, Greenwich Gold has been blended with wheatfeed and marketed under the name Amyplys as a moist concentrate feed for ruminants.

In recent months, the crude protein level in the co-product has been reduced from 28 to about 25 per cent of the dry matter (A van Houte, Amylum; personal communication) but, at that level, Greenwich Gold remains a valuable protein source. With a typical crude fibre level of only 3-4 per cent (in the dry matter), this feed is also a highly digestible energy source. Although this co-product is of cereal origin, the extraction of starch and gluten, together with the proliferation of yeast during fermentation, improves both the quantity and the quality of its protein. Compared to the dry matter of wheat, Greenwich Gold has twice the protein level but almost three times the lysine content.

The gross energy value of the co-product, and consequently the DE value, has proved difficult to determine. A GE value of 21.25 MJ/kg DM was determined at the Institute for Animal Science and Health, Lelystad (Versteegh et al., 2000) and, together with a measure energy digestibility of 89.6 per cent, this indicated a DE value of 19.0 MJ/kg DM. These values seem high in relation to the chemical composition of the feed. Three GE determinations carried out on freeze dried material in the SAC laboratory at Aberdeen gave consistently lower values of between 18.37 and 18.54 MJ.kg DM which, at an assumed digestibility of 90 per cent, would indicate a DE value of 16.6. However, these values relate to samples that may not have been wholly dry. Unfortunately, further oven drying of the freeze-dried material appeared to result in some pyrolysis, with a loss of dry matter as well as water. Thus there are indications that the GE value may be higher then 18.5, but lower than 21.2 MJ/kg DM. Since recent developments have also led to an increase in the fat content, it seems likely that the commercially adopted DE value of 16.5 MJ/kg DM (James & Son, 1999) may represent an understatement of the energy value of this feed for the pig. Further attempts to establish the GE value of the currently available co-product would seem to be justified.

The composition of Greenwich Gold corresponds quite closely to the nutritional objectives of many pig farmers, who prefer a diet with a dry matter content of around 20 per cent, a lysine content of 1-1.2 per cent and a good balance of digestible energy. The specification of Greenwich Gold is very much in-line with this dietary objective, and the consistent composition of the co-product is an aid to diet formulation. An additional advantage, not found in all liquid feeds, is the lack of any tendency to separate into different layers. Thus pig producers can have confidence that they are feeding the intended ration, without any need to stir the product prior to mixing.

Apart from its energy value, the lactic acid content of Greenwich Gold may be of additional value in promoting conditions within the hind gut which favour the development of beneficial micro-organisms, and discourage the growth of undesirable organisms. In a laboratory study at SAC, Aberdeen (I Murray, private communication), the number of viable E. coli inoculated into Greenwich Gold fell one hundred fold within five days, while the number of lactobacilli increased. Such a dramatic effect on gut Microflora may have significant implications for pig health, particularly at a time when the use of in-feed antibiotics is being discouraged.

The mineral content of extracted feeds is always a concern for pig farmers, because the extraction of organic constituents results in an increase in the mineral content of the residue. Of greatest interest are the electrolyte levels, and especially sodium, since the pig has a poor tolerance of high sodium diets. Greenwich Gold has been of particular concern because not only is the sodium concentration increased during wheat processing, but additional sodium, as sodium hydroxide, may be added prior to the fermentation stage in order to adjust the pH. In the past, the sodium content of this co-product has varied widely, but the manufacturers are now well aware of this potential danger and have instituted control procedures and daily monitoring. The high sodium levels seen previously have been avoided and, over the last year, the sodium content has averaged 1.2 per cent of the dry matter (A van Houte, Amylum; personal communication).

A relatively new development, which further reduces the risk of salt toxicity for the pig, is the implementation of welfare codes on UK farms that include the provision of clean water to pigs at all times – including those offered a liquid diet. Although this development is welcome, high sodium levels remain undesirable because of their potentially depressing effect on feed intake, and the stimulation of greater volumes of effluent/pig slurry. Thus the avoidance of a high sodium level in the feed is more beneficial to the pig farmer, than the provision of additional drinking water that improves the animals' ability to cope with a higher mineral intake. A recent study at the University of Plymouth (Brooks and Russell, 2000) included the use of Greenwich Gold at up to 30 per cent of the feed dry matter, with satisfactory pig performance being recorded at that level. The highest inclusion rate for Greenwich Gold in that study resulted in a diet with a sodium concentration of 0.37 per cent of the dry matter.

A positive feature of the mineral composition of Greenwich Gold is the relatively high digestibility of the phosphorus fraction for pigs. The phosphorus availability of wheat is reduced by the substantial proportion that is chemically bound to phytic acid – reported to be 60-75 per cent (NRC, 1998). Dutch workers at the ID-Lelystad (Versteegh et al., 2000) have measured a phosphorus digestibility value of 61 per cent for Greenwich Gold, which they contrasted with a typical value of only 26 per cent for wheat grain.

Greenwich Gold ha also been used for ruminants. It is an ideal ration ingredient that mixes readily with other components, reduces dust levels and prevents the separation of costly micro-ingredients. The co-product also has a high nutritional specification for ruminants, which makes it far more than a carrier of other components. In fact it is a potentially valuable source of both energy and protein, and a useful source of phosphorus. However, Greenwich Gold is a liquid feed and a suitable storage tank would be needed on the farm. Compared to molasses, his feed has much higher moisture content and the required storage capacity would need to be three to four times as great. Providing those facilities are available, Greenwich Gold could prove a valuable supplement for low protein forages, and could go a long way towards meeting the protein of dairy heifers and beef cattle.

Roux™

Roux is the proprietary name given to unrefined "glucose" syrup that is produced as a co-product at the Roquette wheat processing factory at Corby. It is a palatable, mid-brown coloured viscous liquid not unlike molasses, but lighter in colour and easier to mix with other feed ingredients. The high dry matter content of Roux, typically 74-78 per cent, is very largely made up of highly available carbohydrates, some of them simple sugars like dextrose and maltose, but there are also more complex oligo- and poly-saccharides.

Apart from the carbohydrates, Roux contains about 5 per cent protein in the dry matter but only traces of oil and fibre. Since it is not the residual product of wheat processing, Roux has relatively low mineral levels, and the typical ash content is only 4 per cent of the dry matter. The digestibility of this syrup may be expected to be high, and a DMD of 93 per cent has been determined in vitro.

Roux is currently being marketed exclusively to the animal feed trade where it is used principally as a pellet binder. Its high energy content makes it suitable for use in a wide range of diets for both ruminant and non-ruminant animals.

Promanna™

Promanna is a dry, high protein co-product from the Roquette wheat processing factory at Corby. It comprises a concentration of the non-vital (non-gluten) wheat proteins that remain after the gluten and wheat germ have been removed. After separation of this fraction, the co-product is flash-dried and marketed as a free-flowing, tan-coloured powder for use as an ingredient of speciality feeds, such as those for the baby pig, and for other classes of animal that require highly digestible protein sources, such as fish and chicks. In the aftermath of the BSE problem, there has been opposition to the use of fish as well as meat meals in ruminant diets, and this has boosted demand for Promanna as a rich protein source of vegetable origin. Research in progress by the company indicates that 60-65 per cent of the crude protein escapes rumen digestion, and provides a good source of amino acids post-ruminally.

The crude protein content of Promanna, at 63 per cent of the dry matter, is in line with the protein content of fishmeal. Although, the protein is clearly of wheat origin, the amino acid composition of this co-product is quite different from that of the unfractionated wheat grain. In particular, the removal of the substantial gluten fraction leaves a protein that is much richer in the crucially important lysine content. Promanna protein contains 6 per cent lysine, whereas wheat protein comprises only 2.8 per cent lysine. Thus Promanna protein is much more in line wit that of fishmeal (lysine 7.5 per cent of protein), and could be considered to be an alternative protein source for non-ruminant animals. Research into the nutritional value of Promanna is being carried out by the Roquette company in France. Early reports indicate that the ileal digestibility of the crude protein is very high – in line with the published value of 91 per cent for wheat flour (Eurolysine, 1988).

In addition to its high protein content, Promanna also includes significant levels of sugar and starch – amounting to a combined total of 20 per cent of the dry matter. Taken together with a low crude fibre content, this would seem to indicate highly digestible carbohydrate, and the energy value of the feed is also boosted by the low ash and oil content of approximately 10 per cent. The company estimates of DEpig and MEruminant at 20.2 and 15.3 MJ/kg DM respectively, may be conservative. If a digestibility of 90 per cent can be established, pig and ruminant values of 20.9 and 17.0 MJ/kg DM may be justifiable.

Abracarb Plus™

Abracarb Plus is a liquid co-product from the Roquette wheat processing factory at Corby. It comprises a blend of two fractions: a "fibrous" material separated during the initial processing of wheat flour together with the residual liquor that remains at the end. This latter fraction, a carbohydrate-rich residue that remains after the extraction of the gluten, germ, other protein components and the larger granule starch, is evaporated to more than 30 per cent dry matter before re-combination with the lower dry matter fibrous material. The dry matter content of both these fractions has been improved in recent months (S Grainger, Roquette; personal communication), and the blended co-product has a typical dry matter content of 23-25 per cent It is marketed to pig farmers on a dry matter basis, with the DM content of every load individually tested by the factory.

The "fibrous" component of this blend is not in reality a high fibre material, having a crude fibre level of only 9 per cent in comparison with a starch content of 40 per cent (both expresses on the dry matter basis). After blending with larger amounts of the residual liquor, the crude fibre content of the resulting Abracarb Plus is only 2-3 per cent (in the DM), which makes this co-product only a little more fibrous than the original wheat grain. This material has a wide ratio of digestible carbohydrate to fibre and may be expected to be highly digestible.

Abracarb Plus has a crude protein level similar to that of the original wheat grain but, without the gluten fraction, the amino acid composition is quite different. The important lysine, threonine and isoleucine contents are all higher, although the methionine and cystine contents are lower. However, in contrast to the residual material left after fermentation and distillation, Abracarb Plus is essentially an energy feed. Its protein content and protein quality are poorer than those of Greenwich Gold and Vitagold, owing to the lack of any yeast activity.

The mineral content of extract feeds is always a concern for pig farmers because the pig has a poor tolerance of high electrolyte levels, particularly in situations where the water supply is intermittent. The concentration of mineral elements increases during wheat processing as other fractions are removed, and additional amounts of sodium are also added in order to provide the optimum pH level for the enzymic conversion of starch into sugar. Thus the sodium level of the co-product is of particular concern – at a typical range of 0.8-1.4 per cent of the dry matter, this may determine the maximum dietary inclusion rate of Abracarb in pig diets, and the sodium content of other ingredients would have to be considered carefully.

C Starch™

C Starch is an occasional co-product of wheat starch processing at the Amylum factory at Greenwich. It is available when gluten and starch extraction is in progress but the adjacent distillery is unable to process the residual material. The consequence is the availability of a liquid feed containing fibre, oil, minerals and soluble protein from the original wheat flour, together with the small granule C starch that would usually be fermented and distilled. The presence of this additional starch, and the absence of any of the yeast debris that would be left after the fermentation process, means that this co-product is essential an energy source. It has a starch content of 80 per cent in the dry matter but only low crude protein content. At 12-13 per cent, C starch™ has a relatively low dry matter content, and a notably low pH value of 3.2.

AmyPlus™

Amyplus is a blend of wheat starch co-products from the Amylum factory at Greenwich. It is produced by the combination of two co-products from opposite ends of the processing chain; comprising the wheatfeed that was separated at the start of the process and Greenwich Gold, which is the residual liquor remaining at the end. Both of these co-products are available separately and are described in other parts of this chapter.

Amyplus is a moist feed with a dry matter content in the region of 54 per cent, and is aimed at the ruminant feed market. The dry matter content distinguishes it from most other moist feeds, which tend to be in the range of 20-35 per cent. However, the nutrient content of this feed suggests that it will be a competitor with other moist feeds such as brewers' grains. While the ME value and protein content of Amyplus are in line with brewers' grains, the source of the energy is different – Amyplus has higher contents of sugar and starch but a much lower oil content. Thus this co-product blend may be expected to promote greater synthesis of microbial protein in the rumen, but the lower oil content may mean less efficient use of energy at tissue level.

Amyplus is recommended for all ruminant animals. It has a much lower fibre content than brewers' grains and could not be expected to have a roughage sparing effect. Thus Amyplus should be considered as part of the concentrate part of the ration. If it is to be used as a partial replacement for brewers' grains, Amyplus would typically be fed at 4 kg per cay to dairy cows, though higher amounts may be possible. It could also be fed to beef cattle at up to 30 per cent of the diet. For sheep, Amyplus could be used in place of at least part of the diet, though additional calcium would need to be provided to balance the significant phosphorus level.

The magnesium content of this feed, at more than 0.6 per cent, would represent a useful supplement for lactating ewes on spring pastures where there is a risk of hypomagnesaemic tetany.

It is difficult to exclude air from a solid feed of this dry matter content, and good storage is an essential requirement. The feed needs to be ensiled and sheeted effectively even for short storage periods.

12 Other Food Industry Co-products

Potentially, there are many food processing operations that could give rise to products that fail to meet the required specification for a food product. As discussed earlier, such materials are of the highest quality, and they have a provenance and traceability that would satisfy any food safety inspection. Whether or not these materials are recycled into the food chain by their use as animal feeds often depends on their relative value on the farm, and on there being sufficient volume to justify a collection and delivery service by the feed merchant.

Some examples are identified in this chapter. It should be noted, however, that there is relatively little recorded experience and practically no independent evaluation. Some materials have been fed for a number of years, however, and a reasonable estimate of their value can often be made from a chemical analysis. An assessment of each source is probably advisable, because proprietary names do not always indicate the type of material that may be on offer. Foods with similar names may be quite dissimilar in composition, and it is well to be reminded of "rice polishings" and "polished rice", which are completely different fractions, not alternative names. The measurement of dry matter content is of prime importance, and the character of the dry matter can usually be determined by an assessment of oil, ash and fibre. Protein content and quality are obviously relevant, particularly if the material is being considered for non-ruminants.

Chocolate confectionery

Chocolate is available as an animal feed ingredient from all the major chocolate manufacturers and some of the smaller ones too. It is almost never available as pure chocolate but includes a wide range of hard and soft centres, nuts, dried fruit and caramel. In many instances, the co-product represents the finished product that is rejected in the factory because of its non-standard appearance. Often such material will be rejected as non-standard at the start and the end of a production run, and during factory breakdowns. New products will be completely rejected until the factory can consistently produce the required item to the highest standards. Easter and Christmas production peaks also lead to an increased availability of feed chocolate.

An estimated 20,000 tonnes of chocolate enters the animal feed chain each year (P. Latham, SugaRich; personal communication). The majority of this is handled in solid form and processed before inclusion in blended feeds for ruminants and pigs. The most obvious requirement is the removal of the foil, paper and plastic wrappings, and these are typically aspirated away after the co-product chocolate has been put through a machine like a large-scale grater. Perhaps a quarter of the chocolate supply is heated, diluted with water and blended with other confectionery items that have also been converted to a liquid form. This blended confectionery co-product is then marketed to pig farms as **liquid chocolate**. The liquid co-product is supplied by bulk tanker at a warm temperature, and some form of agitation is recommended for the farm storage tank in order to maintain the consistency of the material.

Liquid chocolate is a highly palatable, energy-rich feed material, containing high levels of sugar and vegetable oil. Other nutritional components are present in only low concentration, and supplementation with protein, minerals and vitamins will be needed to produce a balanced ration. The presence of theobromine in chocolate is usually noted. This is an alkaloid of similar chemical structure to the caffeine found in coffee, tea and to a small extent in cocoa. Although theobromine is not regarded as a stimulant, it does have some diuretic effects (McGee, 1984) and it is a prohibited substance in the horse racing world. For pigs, its dilution in chocolate manufacture, further dilution in the manufacture of chocolate confectionary (by mixing with sugar, milk and non-cocoa fats) and ultimate dilution in the pigs' diet (where liquid chocolate is unlikely to be more than a minor (5-10 per cent) ingredient), implies that the theobromine content is unlikely to be of significance.

Stockfeed fruit juice

Canned fruit are used for a variety of confectionary purposes in the food industry. Decanning and removal of the fruit leaves a substantial volume of the juice, and this becomes available for pig feeding. Potentially, a range of fruits may be possible but two types of juice have been identified; viz: "sweet" and "savoury" – as represented by the juice from canned pineapple and tomatoes.

Both types are liquids with a high moisture content but, in the commercially available material, the juice from canned tomatoes has only half the dry matter content of the sweet type. Undoubtedly, this discrepancy is a function of the sugar content of the juice; sweet juice has 84 per cent sugar, and savoury juice 52 per cent sugar in the dry matter. The juice from canned tomatoes consequently has more protein and fibre than the sweet juice. Without knowing the ash and oil content of the savoury juice, one can make only a tentative estimate of the DE value of this material for pigs. The unknown fraction of the sweet juice is less significant . It may be expected that both juices would have a DE value of the order of 16 MJ/kg DM.

The commercially available stockfeed fruit juice is a 60:40 blend of these two juices – a form of sweet and sour sauce for inclusion in the pig diet. It may be regarded as a highly digestible energy source that needs considerable supplementation with good quality protein and adequate amounts of minerals and vitamins.

Jam

Jam is a form of preserved fruit in which the sugar content is increased to the point where the food is microbiologically stable. The sugar concentration results in a high osmotic pressure and this inactivates any microbe that comes into contact with the jam. A combination of natural and added pectin binds the sugar and fruit into a gel, and the water content of approximately 30 per cent is similarly bound. Fruit contains only small amounts of protein and ash and only traces of oil, and since these proportions are substantially diluted by the addition of sugar or glucose syrup, jam may be regarded simply as an energy source. McCance and Widdowson (1991) list the product as 99.1 per cent carbohydrate and 0.9 per cent protein.

The jam that becomes available as a feed contains a significant proportion of plant washings, and is provided in liquid form. The additional water not only dilutes the nutrient content of the jam but it also reduces its microbiological stability. As with all moist feeds, the critical question concerns the dry matter content and this may be expected to be variable. However, the determination of the dry matter content of jam is not simple since, in the oven, the sugar begins to caramelise and lose dry matter whilst water is still present. An alternative solution may be to use a refractometer to measure the sugar concentration and use this to predict dry matter content. Feed loads of jam tested in this way have shown sugar levels of 8-15 per cent, and this equates to approximately 13-24 per cent dry matter.

The provision of clean storage tanks and delivery pipes on the farm will help to reduce fermentation and the loss of dry matter, though some surface moulding may be expected, particularly at lower dry matter content.

Tables of Nutritive Value and Chemical Composition

Typical values are provided from both commercial and scientific sources

In some cases a range of values is provided to give an indication of variability

Co-products of Apple Processing (Chapter 2)

Apart from DM, all values are expressed on DM basis

Feed material	*Cider Apple Pomace*	*Culinary Apple Pomace*
Maximum allowance ruminants*	15	15
Maximum allowance pigs*	NA	NA
Dry matter %	20-28	19†
Ruminant ME (estimated) MJ/kg	9-11.7	11
Ruminant FME (estimated) MJ/kg	7.5-10.2	9.4
Pig DE (estimated) MJ/kg	NA	NA
Crude protein %	6.7	4.5
Oil (A) %	2.3-3.2	
Oil (B) %	2.0-3.3	
Starch %	2-7	
Ash %	2.3	3
NDF %	43-56	34
Crude fibre %	18-25	13.6
Calcium %	0.16	0.2
Phosphorus %	0.14	0.01
Magnesium %	0.06	0.1
Sodium %	0.02	
Potassium %	0.68	
pH	3.5	3.5
Alcohol %		5
Lactic acid %		1.1

Sources: Trident Feeds, Alibes *et al.*, 1984; Givens & Barber, 1987; MAFF, 1990

*The maximum allowances are expressed as % of the dietary dry matter. They aim to represent sensible limits in practical rations. Other dietary factors may necessitate a lowering of the suggested amounts. The allowance refers to dairy cows and higher amounts may be possible in beef cattle diets.

†Value claimed by Trident Feeds, who market the co-product.

Bakery Co-products (Chapter 3)

Apart from DM, all values are expressed on DM basis

Feed material	*Bread*	*Cake*[†]	*Biscuit meal*	*Breakfast cereal*
Maximum allowance ruminants*	20	10[#]	10	20
Maximum allowance pigs*	35	10[#]	10[#]	35
Dry matter %	65 or 90	68-87	90-94	90-93
Ruminant ME (estimated) MJ/kg	14.0	14-19	15.2-17.4	13.5
Ruminant FME (estimated) MJ/kg	13.1	8.7-12.2	10.0-12.3	13.0
Pig DE (estimated) MJ/kg	16.6	18-23	18.8-21.5	16.5
Crude protein %	14.0	5.2-14.7	5-10	9-13
Oil (B) %	3.0	6-35	10-25	0.8-2.5
Starch %	70-73	18-32	21-76[‡]	73
Total Sugars%	3-4	35-63	2.5-45[‡]	9
Ash %	2.8		4.5	2.5
Crude fibre %	2.4		1-3	3.5
NCGD%	94.0			
Calcium %	0.17	.05-.11	0.3	0.1
Phosphorus %	0.15	.13-.29	0.2	0.25
Magnesium %	0.04	.01-.02	0.1	
Sodium %	0.35	.12-.56	0.4	0.2
Potassium %	0.17	.07-.19	0.4	
Lysine %	0.35			
Methionine %	0.23			
Cystine %	0.37			
Threonine %	0.42			
Isoleucine %	0.57			
Tryptophan %	0.18			
pH				

Sources: KW; McCance & Widdowson, 1991; ADAS, 1991c; Vitec-3, SugaRich

*The maximum allowances are expressed as % of the dietary dry matter. They aim to represent sensible limits in practical rations. Other dietary factors, such as the sugar, starch or oil content of other feeds, may necessitate a lowering of the suggested amounts.

The pig allowances refer to fattening pigs and higher amounts may be possible in sow diets.

[†]All values refer to sponge cake - range of 5 types (Source: McCance & Widdowson)

[‡]These values refer to individual biscuit types not to biscuit meal (Source: McCance & Widdowson)

[#]Suggested allowance refers to high fat types - larger amounts of lower fat cake and biscuits may be possible.

Malting, Brewing and Vinegar Co-products (Chapter 4)
Apart from DM, all values are expressed on DM basis

Feed material	*Malt Powder*[§]	*Malt Screenings*	*Malt Culms*	*MRP*	*Brewers' Grains*[×]	*Mash Filter Grains*[×]
Maximum allowance ruminants*	10	15	10	12	30	30
Maximum allowance pigs*	10	15	5	5	NA	NA
Dry matter %	88-95	88	92-96	90	18-25	24-30
Ruminant ME (estimated) MJ/kg	8-13	13.0	11.1	11.5	11.1-12.5[#]	12.0-12.7[#]
Ruminant FME (estimated) MJ/kg	7.5-11.7	11.9	10.2	10.8	8.2-9.6	8.9-9.6
Pig DE (estimated) MJ/kg	8.5-15	15	11	11	NA	NA
Crude protein %	6-17	10-13	20-32	23	19-31	20-25
Oil (B) %	1.5-4.5	3.8	2.9	2.5	8.5-11.0	9-12
Starch %			3-19	15.8	2-8	4-11
Total Sugars%			7-25	3.5	<0.1-0.6	<0.1-3.0
Ash %	2-11	2.6	5.7	6.2	3.5-9.0	4.8
NDF %	12-71	16-21	48.4	48.0	50-64	49-57
ADF %					19-24	21.0
Crude fibre %	4-24	5.9	9-14			
NCGD %	56-89	85-88	67-79	68-78	56-68	60-66
Calcium %			0.24	0.25	0.25-0.55	0.37
Phosphorus %			0.71	0.57	0.6-0.7	0.60
Magnesium %			0.16	0.16	0.18-0.23	0.22
Sodium %			0.04	0.14	0.02	0.02
Potassium %			1.6	0.8	0.06	0.14
Lysine %						
Methionine %						
Cystine %						
Threonine %						
Isoleucine %						
pH					4-5	4.7-5.2
Alcohol %						
Acetic acid %						

MRP represents malt residual pellets, and LME is Liquid Malt Extract
Sources: James & Son; Lonsdale, 1989; Ewing, 1997; MAFF, 1990; Rowett, 1994b; Martin, 1982
*The maximum allowances are expressed as % of the dietary dry matter. They aim to represent sensible limits in practical rations. Other dietary factors, such as the oil content of other feeds, may necessitate a lowering of the suggested amounts.
The pig allowances refer to fattening pigs and higher amounts may be possible in sow diets.
† The dry matter content includes 3.5-8% alcohol by volume - equivalent to 2.8-6.3% w/w.
‡ The DE value relates to beer with an alcohol content of 4.5% v/v.
× The values for brewers', mash filter, vinegar and malt extract grains represent separate datasets.

Vinegar Grains[×]	*Malt Extract Grains*[×]	*Black Grains*[††]	*Grains Pressings*	*Brewers' Yeast*	*Beer*	*Vinegar Still Bottoms*	*LME*
30	30		NA	10	NA	5	10
NA	NA		20	20	10	5	20
21.5	21-27	27.40	9.0	12-16	5.5-9.5[†]	67.4	76-78
11.7-12.7[#]	11.9-13.3[#]	9-10.7	NA	13.5	NA	14	
8.6-9.6	9.3-10.7	7-8.7	NA	12.5	NA	14	
NA	NA	NA	18	18	21[‡]	17.3	
21.7	21-24	21-29	38.0	36-50	4-7	33.1	6.4
8.6-12.7	7.5-10	6-8	7.9	2.5-4.5		0.15	0
7-13	6-18	6-11	23.0	2-20		2.0	<0.1
0.5	0.1-4.0	0-1	1.4	0.1-2.0	10-18	2.8	91[§§]
4.2	4.6	3-17	3.0	6-10	9-16	10.8	1.6
50-59	43-57	45-55	17	3-10		0.1	0
21.8	21.6						
63-65	63-74	33-49	90	88-93		89.1	
0.23	0.36		0.15	0.2-0.3		0.07	0.02
0.66	0.60		0.42	1.5-2.0		1.65	0.21
0.25	0.21		0.10	0.2-0.3		0.40	0.06
0.02	0.03		0.04	.02-.10		0.13	0.03
0.11	0.34		0.67	2.0-2.6		3.58	0.56
			1.30	2.77			
				0.61			
				0.50			
				1.91			
				1.85			
4.0-4.2	3.8-5.2					4.2	3.5
					48-68		
						16.4	

[#] The ME values have been calculated from the NDF values (Rowett 1984b)

[§] Malt Powder includes barley screenings and it is a variable feed material that covers the range from barley flour to thin grain mixed with straw and chaff.

[††]The ME values have been calculated from the NCGD values (Rowett NCD equation;1984b)

[§§]This fraction includes both simple and more complex sugars

Co-products of Citrus and Tropical Fruit Processing (Chapter 5)

Apart from DM, all values are expressed on DM basis

Feed material	*Citrus Pulp*†	*Orange Pulp*†	*Lemon Pulp*†	*Citrus Molasses*	*Fruit Salad*
Maximum allowance ruminants*	12	12	10	10	15
Maximum allowance pigs*	5	5	NA	10	10
Dry matter %	17-24	17-24	16-19	71.0	8.7-9.5
Ruminant ME (estimated) MJ/kg	13.6	13.6	13.0	11.3	11
Ruminant (FME) (estimated) MJ/kg	12.8	12.8	12.7	12.8	9.5
Pig DE (estimated) MJ/kg	15.0	15.0	NA	12.9	13
Crude protein %	6.8-9.7	7.9	6.8	5.8	11-13
Oil (B) %	1.1-3.7	1.1-3.5	1.0	0.3	5.2
Starch %	0.1-8.8	2.5-21	5.0		11.5
Total Sugars%	1.4-28	2-28	10-14	63.4	1.0
Ash %	4.3	3.9	5.0	6.6	5.3-10.7
NDF %	19-26	20.8	28.7	0.0	33
Crude fibre %				0.0	18.6
NCGD%	90-95	92.3	91.0		74-83
Calcium %	0.72	0.74	0.98	1.13‡	0.48
Phosphorus %	0.12	0.12	0.10	0.08	0.44
Magnesium %	0.10	0.10	0.09	0.14	0.25
Sodium %	0.03-0.16	0.04	0.11	0.42	0.29
Potassium %	0.7-1.3	0.74	0.76	1.55	3.18
pH	3.5-4.1	3.5-4.1	4.0	5	3.6

Sources: James & Son; Hendrickson and Kesterson, 1965; NRC, 1989

*The maximum allowances are expressed as % of the dietary dry matter. They aim to represent sensible limits in practical rations. Other dietary factors may necessitate a lowering of the suggested amounts.

The pig allowances refer to fattening pigs and higher amounts may be possible in sow diets.

The rumimant allowances refer to dairy cows and higher amounts may be possible in beef cattle diets.

† Citrus pulp information relates largely or wholly to orange pulp; the orange and lemon pulp data come from more limited data sources.

‡The calcium value of the molasses includes additional calcium added during processing.

Distillery Co-products (Chapter 6)

Apart from DM, all values are expressed on DM basis

Feed material	*Draff*	*PAS*	*Supergrains*
Maximum allowance ruminants*	30	10	30
Maximum allowance pigs*	NA	30	NA
Dry matter %	20-26	30-50	25.0
Ruminant ME (estimated) MJ/kg	11.0	15.6	14.1
Ruminant FME (estimated) MJ/kg	7.8	14.7	10.9
Pig DE (estimated) MJ/kg	NA	16.7	NA
Crude protein %	20-23	34-38	29.0
Oil (A) %		2.0	
Oil (B) %	9-13	3.0	9-12
Starch %	0-5	1.3	5.0
Total Sugars% (Luff Schoorl)	0.2	2.5	0.4
Ash %	3.5	9.5-10.5	2.5
NDF %	60-66	0.6	53-64
Crude fibre %	17.0	0.2	16-19
NCGD%	48-58		63-68
Calcium %	0.20	0.14-0.20	0.14
Phosphorus %	0.50	1.6-2.2	0.33
Magnesium %	0.20	0.65	0.07
Sodium %	0.01	0.10-0.15	0.01
Potassium %	0.05	2.1-2.3	0.19
Copper mg/kg	8-18	60-180	30-90
Lysine %		2.1[#]	
Methionine %		0.35[#]	
Cystine %		0.7[#]	
Threonine %		1.9[#]	
Isoleucine %		1.3[#]	
Tryptophan %			
pH		3.5-3.8	4.1

PAS and ESW are pot ale syrup and evaporated spent wash respectively; LL Supers and LL Gold are co-products from the Loch Lomond Distillery

Sources: Gizzi, 2001; MAFF, 1990; Rowett, 1984; RRI, Trident Feeds, WPSA (1992)

*The maximum allowances are expressed as % of the dietary dry matter. They aim to represent sensible limits in practical rations. Other dietary factors, such as the sugar, starch or oil content of other feeds, may necessitate a lowering of the suggested amounts. The pig allowances refer to fattening pigs and higher amounts may be possible in sow diets.

[†] Adjusted to 25% protein

LL Supers	*LL Gold*	*ESW*[×]	*Vitagold*	*Distillers Malt*	*Distillers Wheat*	*Distillers Maize*
30	10	10	30	35	35	35
NA	30	30	NA	NA	NA	NA
27.0	30-32	34-40	34	90	90	90
13.6	14.0	14.0	14.1	12.6	13.3	14.9
11.2	13.0	13.0	11.4	9.9	11.1	11.5
NA	16.8	16.8	NA	NA	NA	NA
31.0	23-27	23-27	37	27	32	29
			9.3	7.5	6	10
6-9	3.5	3.5		9.0	7.5	11.5
7.0	1-12	1-12		2.5	4.5	2.5
2.0	4-15	4-15		4.0	6.0	1-10
2.8	8	8	3	6.0	5.3	4.5
38-51	3.5	3.5	50	42	23-46[‡]	23-51[‡]
			8.6	12	7-10	9
67-71	92	92	72	67-72	77-88	82
0.11	0.15	0.15	0.1	0.16	0.18	0.14
0.53	1.2-1.5	1.2-1.5	0.5	0.96	0.88	0.84
0.08	0.47	0.47	0.07	0.32	0.28	0.32
0.02	0.12	0.12	0.02	0.11	0.31	0.12
0.30	2.5	2.5	0.3	0.99	1.22	0.95
4-13	45-70	45-70	31-85	30-55	10-120	10-120
	0.80[†]	0.80[†]		1.00	0.65	0.61
	0.45[†]	0.45[†]		0.38	0.42	0.51
	0.45[†]	0.45[†]		0.49	0.48	0.39
	0.74[†]	0.74[†]		1.04	1.00	1.03
	0.83[†]	0.83[†]		1.06	1.15	0.98
		0.22	0.29	0.27		
4.0	4.0	4.0				

[‡]Range of recorded values may include analytical problems

[×]Values for ESW are assumed to be similar to those of LL Gold

[#] Adjusted to 34% protein

Co-products of Maize Fractionation (Chapter 7)

Apart from DM, all values are expressed on DM basis

Feed material	*Maize Screenings*	*CCSL*[†]	*Maize fibre*
Maximum allowance ruminants*	20	7	25
Maximum allowance pigs*	30	5	15-20
Dry matter %	87.0	45-50	35-40
Ruminant ME (estimated) MJ/kg	13.7	11.5-13.0	12.2-14.5
Ruminant FME (estimated) MJ/kg	12.8	11.2-12.7	11.6-13.3
Pig DE (estimated) MJ/kg	16.0	14.0-15.9	12.7
Crude protein %	9.2	40-45	11-21
Oil (B) %	3.0	0.1-2.0	2-4
Starch %	72	2-5	10-25
Sugars%	1.0	2-24	1.5
Lactic acid %		7-15	
Ash %	0.6	9-22	1-7
NDF %	16.0	0.0	36-60
ADF %		0.0	12-18
Crude fibre %	3.5	0.0	10-15
NCGD %	88.0	78-83	57-67
Calcium %	0.10	.04-.10	0.03
Phosphorus %	0.30	1.5-3.0	0.1-1.1
Magnesium %	0.10	1-1.2	0.1-0.5
Sodium %	0.10	0.7-0.8	0.01-0.6
Potassium %	0.40	3-5	0.1-1.8
Lysine %	0.24	1.43	
Methionine %	0.21	0.76	
Cystine %	0.22	0.92	
Threonine %	0.34	1.43	
Isoleucine %	0.33	1.22	
Tryptophan %	0.09		
Xanthophyll mg/kg			
pH		4.0	

CCSL, MGF and MGM are concentrated corn steep liquor, maize gluten feed and maize gluten meal respectively
MGermM is maize germ meal
Sources: Amylum; Cargill; Cerestar; Ewing, 1997; Lonsdale, 1989; MAFF, 1990; WPSA, 1992
*The maximum allowances are expressed as % of the dietary dry matter. They aim to represent sensible limits in practical rations. Other dietary factors may necessitate a lowering of the suggested amounts. The pig allowances refer to fattening pigs and

MGF	*MGM*	MGermM
25-30	5	15
20	5	15
86-91	89	88
11.3-14.2	14.5	13.6
10.4-12.2	12.6	11.9
13-13.5	18.7	14.7
20-25	66	25-26
3-7	6	3.4
9-28	15.5	23
1-3	1	7
2-4		
4-10	2	3
33-45	7.5	37
8-15		15
6-10	1.5	10
69-80	95	80
0.25	0.05	0.1
0.6-1.1	0.3	0.7
0.2-0.5	0.05	0.15
0.03-0.5	0.1	0.05
0.7-1.9	0.1	0.3
0.68	1.14	0.96
0.38	1.61	0.48
0.47	1.18	0.54
0.78	2.27	1.01
0.71	2.76	0.86
0.13	0.44	0.25
12-40	150-300	3

higher amounts may be possible in sow diets. The ruminant allowances refer to dairy cows and higher amounts may be possible in beef cattle diets.

† Concentrated corn steep liquor (CCSL) appears to be a very variable material; the amino acid levels refer to a low ash product produced by Cerestar

‡ Adjusted to 67 per cent protein.

× The allowances relate to the UK-produced co-product; imported material may have a higher oil content and may need to be fed in more restricted amounts.

Milk Processing Co-products (Chapter 8)

Apart from DM, all values are expressed on DM basis

Feed material	*Whey*	*Whey Concentrate*	*Whey Permeate/ Concentrated Whey Permeate*	*Delactosed Whey*	*Yoghurt*	*Ice Cream*
Maximum allowance ruminants*	15	15	15	15	NA	NA
Maximum allowance pigs*	20	25	20	25	20	10
Dry matter %	5-7	30-50	18, 25 or 45	38-45	5-9	12-15
Ruminant ME (estimated) MJ/kg	13.5	13.5	11.6	11.2	NA	NA
Ruminant FME (estimated) MJ/kg	11.4†	13.2	11.5	10.8	NA	NA
Pig DE (estimated) MJ/kg	16.5	16.5	15.0	13.8	14.7-16.8	20-21
Crude protein %	13-15	12.4	3.8	24	18-34	9-10
Oil (B) %	1.0	1.4	0.2	1.2	1.5-8.0	20-25
Total Sugars%	75	75	83	55-60	50-75	60
Ash %	10.0	7.4	11.0	16	5	4
NDF %	0.0	0.0	0.0	0	0-0.7	trace
Calcium %	1.00	0.84	0.86	2.1	0.6-1.3	0.34
Phosphorus %	0.75	0.72	0.66	1.4	0.6-1.0	0.26
Magnesium %	0.08	0.07	0.07	0.26		0.03
Sodium %	1.0-1.3	0.69	1.00	1.9	0.35-0.65	0.26
Potassium %	1.20	1.00	2.10	4.9		0.41
Lysine %	0.94	0.94	0.18	2.06	0.8-1.6	0.63
Methionine %	0.24	0.24	0.03	0.53	0.2-0.4	0.16
Cystine %	0.24	0.24	0.04	0.63	0.2-0.4	0.16
Threonine %	0.70	0.70	0.14	1.60	0.6-1.2	0.47
Isoleucine %	0.66	0.66	0.17	1.58	0.56-1.12	0.44
Tryptophan %	0.21	0.21	0.03	0.42	0.18-0.36	0.14
pH	3.25-4.0					

Sources: Taymix; Wheyfeed; WPSA, 1992; Ling, 1956; McCance & Widdowson, 1991; NDC, 2000; NRC, 1998

*The maximum allowances are expressed as % of the dietary dry matter. They aim to represent sensible limits in practical rations. Other dietary factors, such as the sodium content of other feeds, may necessitate a lowering of the suggested amounts. The pig allowances refer to fattening pigs and higher amounts may be possible in sow diets. The ruminant allowances refer to dairy cows and higher amounts may be possible in beef cattle diets.

†The FME value includes an adjustment for the lactic acid concentration

Potato Co-products (Chapter 9)

Apart from DM, all values are expressed on DM basis

Feed material	*Potato Feed*	*Potato Feed Permeate*	*Potato Feed Solids*	*Abraded Peel*	*Potato Skin*	*Off-cuts/ Canning potatoes*	*Potato Slice*	*Peel & Trim*
Cooked/uncooked	C	C	C	U	C	U	U	U
Maximum allowance ruminants*	15-20	20	15	10-15	10	20	15-20	20
Maximum allowance pigs*	25	30	NA	NA	NA	5	NA	NA
Dry matter %	10-14	9-13	11-15	7.0	10.6	21.1	17.4	32.4
Ruminant ME (estimated) MJ/kg	11.7	12.6	10.4	8.5-10.7	5.5	13.3	11.1	12.6
Ruminant FME (estimated) MJ/kg	11.2	12.3	9.8	7.7-9.9	4.3	13.0	10.5	11.5
Pig DE (estimated) MJ/kg	14.0	14.3	NA	NA	NA	16	NA	NA
Crude protein %	17.0	18.5	17.0	11.4	18.4	7.6	9.3	6
Oil (B) %	1.6	1.0	2.0	2.6	3.9	0.9	1.9	3.77
Starch %	36.6	43.0	30.0	24-38	4.2	75.3	60.1	66.9
Sugars%	1.6	1.5	1.5	0.2	0.2	1	0.5	0.14
Ash %	9.3	10.0	10.0	8.4	6.6	3.2	6.1	10.5
NDF %	22.1	11.0	33.0	32-52	70.1	10	25.5	18.5
Crude fibre %	9.6							
NCGD %	80.5	88.0	71.0	55-73	32.4	93.6	75.7	83
Calcium %	0.24				0.85	0.16	0.22	0.2
Phosphorus %	0.23				0.09	0.15	0.14	0.11
Magnesium %	0.19				0.10	0.06	0.05	0.05
Sodium %	0.02				0.03	0.04	0.06	0.06
Potassium %	3.58				1.61	0.99	0.75	0.46
Lysine %	0.74							
Methionine %	0.23							
Cystine %	0.35							
Threonine %	0.48							
Isoleucine %	0.52							
Tryptophan %	0.47							
pH	4.4							

Sources: James & Son, Frito Lay
*The maximum allowances are expressed as % of the dietary dry matter. They aim to represent sensible limits in practical rations. Other dietary factors, such as the oil content of other feeds, may necessitate a lowering of the suggested amounts.
The pig allowances refer to fattening pigs and higher amounts may be possible in sow diets.
The rumimant allowances refer to dairy cows and higher amounts may be possible in beef cattle diets.

Potato Mash	*Potato Flake*	*Potato Chips*	*Southern Fries*	*Hash Browns*	*Jacket Wedges*	*Potato Crisps*	*Potato Starch*	*PPP*	*PPF*	*Primary Sludge*
C	C	C	C	C	C	C	U	C + U	C + U	U
20	25	10-15	10	10	15	6	5	20	20	15
30	30	10-15	10	10	15	6	30	25	20	25
21.8	90.9	34.0	39.9	38.6	32.4	97.3	60	17-22	14-17	49.7
13.5	13.4	14.8-16.4	16.4	16.1	15.0	22.8	13.9	13.0	12.4	12.8
13.1	13.1	11.9-10.2	10.4	10.6	11.6	11.7	13.8	12.5	12.0	12.7
16.4	16.2	18.3	18.0	17.4	17.8	22.7	17.4	14.9	14.1	15.6
8.1	8.5	6.9	6.0	7.1	7.6	5.4	1	10.8	13.8	1.8
1.4	1.0	10-20	20.3	18.8	11.7	37.6	<0.2	1.8	1.4	0.3
69.0	78.9	65.3	42.9	52.9	60.7	54.8	97.7	65	42.7	92.9
0.6	1.2	0.6	0.3	0.1	0.3			1.3		
3.0	4.5	3.2	3.7	6.2	3.6	3.2	0.30	5.7	7.5	3.6
7.4	5.4	5.8	14.1	10.8	6.5	0.4	0.7	12.5	15.6	6.3
2.5		1.8								
93.9	94.0	95.0	95.5	91.6	94.2	95.8		89.9	86.4	
0.11	0.05	0.04	0.06	0.02	0.04			0.26		
0.15	0.15	0.26	0.35	0.34	0.24			0.26		
0.06	0.05	0.06	0.04	0.04	0.06			0.07		
0.10	0.03	0.15	0.73	1.75	0.51			0.06		
0.86	0.96	1.14	0.68	0.87	1.26			0.41		

Sugar Beet Co-products (Chapter 10)

Apart from DM, all values are expressed on DM basis

Feed material	*Sugar Beet Tails*	*DMSBF*	*PressedBeet Pulp*	*Molasses*	*CMS*[†]
Cooked/uncooked	U	C	C	C	C
Maximum allowance ruminants*	15	30	30	10	5
Maximum allowance pigs*	10	20	20	10	2-3
Dry matter %	9-15	88.0	21-30	74-78	56-60
Ruminant ME (estimated) MJ/kg	10-11.5	12.5	12.5-13.0	10.3-12.0	9.5
Ruminant FME (estimated) MJ/kg	10-11.5	12.4	12.3-12.8	10.3-12.0	9.5
Pig DE (estimated) MJ/kg	13.0	14.0	12.8	12.7	3.5
Crude protein %	7.5	11.0	10.0	10-14	33-36
Oil (B) %	0.3	0.4	0.7	trace	0-1.5
Starch %		6.5	0.4	0.0	0.0
Sugars%		28.1	4.5	63.0	5.0
Ash %	7-25	8.8	8.2	11.0	26-29
NDF %		32.1	52.4	trace	trace
Crude fibre %	8.0	13.2	19.5	trace	trace
NCGD%		86.0	80.8		
Calcium %		0.76	1.20	0.12	0.5-1.0
Phosphorus %		0.08	0.12	0.04	0.1-0.4
Magnesium %		0.11	0.18	0.01	0.5
Sodium %		0.44	0.05	2.5	3.0
Potassium %		1.82	0.60	5.0	6-9
Lysine %		0.42	0.51		
Methionine %		0.10	0.17		
Cystine %		0.12	0.12		
Threonine %		0.36	0.42		
Isoleucine %		0.34	0.36		
Tryptophan %		0.06	0.08		

DMSBF and CMS are dried molassed sugar beet feed and concentrated molasses solubles respectively.

Sources: Trident Feeds; Quest International; MAFF, 1990; WPSA, 1992; James & Son; United Molasses

*The maximum allowances are expressed as % of the dietary dry matter. They aim to represent sensible limits in practical rations. Other dietary factors, such as the sugar content of other feeds, may necessitate a lowering of the suggested amounts.

The pig allowances refer to fattening pigs and higher amounts may be possible in sow diets.

The ruminant allowances refer to dairy cows and higher amounts may be possible in beef cattle diets.

[†]CMS is also available from imported sources and may be the co-product of other processes -its composition may be different from that shown.

Wheat Starch Co-products (Chapter 11)

Apart from DM, all values are expressed on DM basis

Feed material	*Wheatfeed*	*C Starch*	*G. Gold*
Maximum allowance ruminants*	30	NA	20
Maximum allowance pigs*	20	30	30
Dry matter %	89.0	12-13	21-23
Ruminant ME (estimated) MJ/kg	11.9	NA	13.5
Ruminant FME (estimated) MJ/kg	10.4	NA	11.3
Pig DE (estimated) MJ/kg	11.7	16.0	17.0
Crude protein %	18.0	8.5	25
Oil (B) %	5.0	2-3	6-9
Starch %	15-45	78	3
Sugars%	6.0	2	2-15
Ash %	5.0	2.3	6.5-8.5
NDF %	24-44		8-12†
Crude fibre %	4-10	7	3
NCGD%	75		85-88†
Calcium %	0.11	0.05	0.19
Phosphorus %	1.05	0.18	0.62
Magnesium %	0.52	0.24	0.15
Sodium %	0.02	0.60	1.30
Potassium %	1.30	0.38	1.37
Lysine %	0.72	0.28	1.05‡
Methionine %	0.27	0.14	.54‡
Cystine %	0.42	0.23	.59‡
Threonine %	0.59	0.25	.87‡
Isoleucine %	0.57	0.34	.96‡
Tryptophan %	0.25	0.11	0.14‡
pH		3.2	3.3-3.8

Sources: James & Son; IDDLO (Lelystad); Amylum; Roquette; MAFF, 1990

*The maximum allowances are expressed as % of the dietary dry matter. They aim to represent sensible limits in practical rations. Other dietary factors, such as the sodium content of other feeds, may necessitate a lowering of the suggested amounts.

The pig allowances refer to fattening pigs and higher amounts may be possible in sow diets.

The ruminant allowances refer to dairy cows and higher amounts may be possible in beef cattle diets.

† Analytical problems have been encountered when samples are pre-dried before analysis

‡ Adjusted to 25% crude protein in the dry matter

Amyplus	*Roux*	*Promanna*	*Abracarb*
30	6	5	NA
NA	10	10	30
55	74-78	90	23-25
12.1	13	16.1	NA
10.4	12.9	13.2	NA
NA	16	20.5	16
21	5	63	10-12
5.8	0.2	10	1.5-1.7
21	}	9	42-55
2.5	} 90	11	21
5.3	4	3	3-8
34		19	7-8
	<0.5	1	2
76	92		87-93
0.12	0.4	0.15	0.2-0.4
0.85	0.3	0.47	0.24-0.40
0.29	0.1	0.03	0.07
0.22	0.8	0.24	0.8-1.4
1.2	0.9	0.22	0.70
		3.74	0.53
		0.64	0.16
		1.14	0.16
		2.05	0.57
		2.04	0.57
		0.28	0.25
	3.5-4.5		3.5-4.5

Other Co-products (Chapter 12)

Apart from DM, all values are expressed on DM basis

Feed material	*Liquid chocolate*†	*Fruit juice*†	*Jam*
Maximum allowance pigs*	10	10	10
Dry matter %	55.0	8	13-24
Pig DE (estimated) MJ/kg	19.5	16	15.2
Crude protein %	4.5	8.6	0.9
Oil (B) %	16.4		Trace
Total Sugars%	70.0	71.0	99
Ash %	0.7		Trace
NDF %		14.7	Trace
Crude fibre %	4.5		
Calcium %	0.18		
Phosphorus %	0.09		
Sodium %	0.11		

Sources: SugaRich; Taymix; McCance & Widdowson, 1991

*The maximum allowances are expressed as % of the dietary dry matter. They aim to represent sensible limits in practical rations. Other dietary factors, such as the oil content of other feeds, may necessitate a lowering of the suggested amounts. The pig allowances refer to fattening pigs and higher amounts may be possible in sow diets.

†The analyses refer to specific sources from SugaRich and Taymix respectively. They are thus examples of these types of co-product and may not be representative of similarly labelled feeds from other sources

References

ABM (1998) Control of ABM's Certification Activities; an Assured British Meat brochure published in 1998, King's Scholar's House, 224-230 Vauxhall Bridge Road, London, England.

ABTA (1999) Code of Practice for the Supply of Moist Feeds; Allied Brewery Traders Association, Wolverhampton, England.

ACCS Assured Combinable Crops Scheme Manual, 3rd Edition 1999-2000, ACCS Registrar, UKFQC, Oxford, England.

ADAS (1970) Brewers' grains for fattening cattle 1969-70; Report of a study at Gleadthorpe Experimental Husbandry Farm, Mansfield, England.

ADAS (1979) Sugar beet pulp by-products; Nutrition Chemistry Feed Evaluation News 10/79, Stratford on Avon, England.

ADAS (1982) Maize gluten feed in diets for fattening pigs; Interim Report No 82/46, Terrington Experimental Husbandry Farm, Kings Lynn, England.

ADAS (1986a) Maize gluten feed; Nutrition Chemistry Feed Evaluation Unit Technical Bulletin 86/12, Stratford on Avon, England.

ADAS (1986b) Potatoes for Livestock, Advisory Leaflet P467.

ADAS (1986c) Beet Molasses; Nutrition Chemistry Feed Evaluation Unit Technical Bulletin 86/6, Stratford on Avon, England.

ADAS (1987) Fresh and ensiled maize fibre; Nutrition Chemistry Feed Evaluation Unit Technical Bulletin No 87/1, Stratford on Avon, England.

ADAS (1988) Whey for pigs; Advisory Leaflet P3149.

ADAS (1989) Tables of Rumen Degradability Values for Ruminant Feedstuffs, ADAS Feed Evaluation Unit, Stratford on Avon, England.

ADAS (1990) Maize gluten (feed) and rapeseed meal as protein supplements to barley or wheat for intensively finished beef cross dairy bulls; High Mowthorpe Experimental Husbandry Farm N/N3/BC03025, Malton, England.

ADAS (1991a) Overheated maize gluten feed; ADAS South West Region Nutrition Chemistry Newsflash, No 9, January 1991.

ADAS (1991b) Feeding molassed sugar beet feed to pregnant ewes; Report of Rosemaund EHF for Trident Feeds, Peterborough, England.

ADAS (1991c) Nutrition Chemistry Feedstuffs Database; compiled by P Clark.

ADAS (1994) A study into the digestibility and metabolisable energy content of liquid potato feed and potato chips; Feed Evaluation Unit Contract Report C004066 for James & Son, Northampton, England.

AFRC (1993) Energy and Protein Requirements of Ruminants; published by CAB International, Wallingford, England.

Alderman G (1987) The nutritive value of wet maize gluten feed; a technical report prepared for Tunnel Refineries of Greenwich, now Amylum UK, Greenwich, London.

Alibes X, Muñoz F and Rodriguez J (1984) Feeding value of apple pomace silage for sheep; Anim. Feed Sci. Technol. **11,** 189-197.

Alibes X, Rodriguez J, Muñoz F and Geria R (1979) Valor alimenticio del ensilado de pulpa de manzana suplementación con distintas fuentes de nitrógeno; IV Jornadas Cientificas de la Sociedad Española de Ovinotecnia, pp 213-223.

Anderson M J, Lamb R C, Mickelsen C H and Wiscombe R L (1974) Feeding liquid whey to dairy cattle; J. Dairy Sci. **57,** 1206.

Anon (1985) Brewery wastes prove excellent pig feed; East Anglian Daily Times, January 19, 1985, page 8.

Anon (1992) Typical composition of feeds for cattle and sheep, 1992-93; Feedstuffs, Special Report 18 May 1992, pp35-40.

Anon (1992b) Encyclopedia of Food Science and Technology; edited by Y H Hui, an Interscience Publication by A Wiley.

Anon (1999) Try dry! Annual update on the potato flake industry, Potato Business World **7** (4), 20-31.

Anon (2001) The UK Pesticide Guide, edited by R Whitehead, published by British Crop Protection Council / CAB International, Bracknell, England .

ARC (1980) The Nutrient Requirements of Pigs; published on behalf of the Agricultural Research Council by CAB, Slough, England.

ARC (1981) The Nutrient Requirements of Ruminant Livestock; published on behalf of the Agricultural Research Council by CAB, Slough, England.

Archer K A, Rogan I M and Bowen R W (1980) A comparison of potatoes with grain sorghum in feedlot diets for production of prime lambs; Aust. J. Exp. Agric. Anim. Husb. 19 (101), pp 679-683.

Askar A and Treptow H (1997) Tropical fruit processing waste management – part I: waste reduction and utilisation; Fruit Processing **7 (9),** 354-359.

Atkins C D, Wiederhold E and Moore E L (1945) Fruit Production Journal **24,** p260.

Austin Rebecca (1996) Supermarkets – join quality scheme; Farmers Weekly, issue of 5 July 1996.

Baker F S Jr (1950) Citrus molasses in a steer fattening ration; Florida Agr. Exp. Sta. Circ. S-22.

Barber J (1998) It pays to pay attention to water-to-feed ratios in liquid diets; Pig World, January 1998, page 41.

Barber R S, Braude R, Mitchell K G and Pitman R J (1978) The nutritive value of liquid whey, either sour or sweet, when given in restricted amounts to the growing pig; Anim. Feed Sci. Technol. **3,** 163-177.

Bayley S (1994) Gin; published by the Gin and Vodka Association to mark its 50th anniversary, Andover, England.

Beames R M and Taylor B S (1991) Liquid feeding of pigs; The University of British Columbia Final Report ARDSA Project 11006, Vancouver, Canada.

Beever D E, Mould F L and Mauricio R M (1999) An assessment of the potential of potato co-products as feedingstuffs for ruminants; Project Report No 111 by the University of Reading Department of Agriculture on behalf of James & Son, Northampton, England.

Beever D E, Sutton J D, Thomson J D, Napper DJ and Gale D L (1988) Comparison of molassed and unmolassed sugar beet feed and barley as energy supplements on nutrient digestion and supply in silage-fed cows; Anim. Prod. **46,** 490 Abstr.

Bennion E B and Bamford G F T (1983) Technology of Cake Making; 2nd Edition, Leonard Hill Books.

Beudeker et al (1990) Baker's yeast; In: Yeast – Biotechnology and Biocatalysis, Eds. Verachtert and De Mot, Marcel Dekker Inc.

BFBi (2000) Code of Practice for the Supply of Moist Co-products; Brewing, Food and Beverage Industry Suppliers Association, Wolverhampton, England.

Bhattacharya A M, Khan T N and Uwayjan M (1975) Dried beet pulp as a sole source of energy in beef and sheep rations; J. Anim. Sci. **41,** 616-621.

Black H, Edwards Sandra, Kay M and Thomas S (1991) Distillery By-Products as Feeds for Livestock: A report made to the Malt Distillers Association of Scotland, Elgin, Scotland.

Blair J (2001) Cider; In: Excellence in Packaging of Beverages; edited by J Browne and E Candy, published by Binsted Group plc.

Blaxter K L and Clapperton J L (1965) Prediction of the amount of methane produced by ruminants; Br. J. Nutr. **19,** 511-522.

BLRA (1994) The Story of Beer; Brewers and Licensed Retailers Association, London, England.

BLRA (2000) Statistical Handbook: A compilation of drinks industry statistics; published by Brewing Publications Ltd, London, England.

Boucqué Ch V, Cottyn B G, Aerts J V and Buysse F X (1976) Dried sugar beet pulp as a high energy feed for beef cattle; Anim. Feed Sci. Technol., **1,** 643-655.

Boucqué Ch V and Fiems L O (1988) Vegetable by-products of agroindustrial origin; Livestock Prod. Sci. **19,** 97-135.

British Potato Council (1999) Potato Consumption and Processing in Great Britain, Oxford, England.

British Sugar (1998) Facts about British Sugar 1998/9, Peterborough, England.

Brooks P H (1999) The multiple benefits of liquid feed for pigs; Feed Tech **3,** 29-31.

Brooks P H and Russell P J (2000) The effect of graded levels of Greenwich Gold on the performance of growing-finishing pigs; University of Plymouth report to Amylum Group, Aalst, Belgium.

Brouwer E (1952) 'On the base excess, the alkali alkalinity, the alkaline earth alkalinity and the mineral ratios in grass and hay with reference to grass tetany and other disorders in cattle'; Brit. Vet. J. **108,** 123.

Brown R S and Eden A Self-feeding of brewers' grains; internal MAFF report.

BSF Sugar Beet Feeds; Badminton Speciality Feeds leaflet, Oakham, England.

Cabezas M T, Hentges J F, Moore J E and Olson J A (1965) J. Anim. Sci. **24,** 57.

Cahill D, McAleese D M and Ruane J B (1966) A comparison of concentrate feeds for intensive beef production; Ir. J. Agric. Res. **5,** 27-41.

Carr J G (1970) Modern methods of cider making; National Association of Cider Makers, London, England.

Castle M E (1972) A comparative study of dried-sugar beet pulp for milk production, J. Agric. Sci Camb. **78,** 371-377.

CEDAR (1995) Mixed forage diets for dairy cows, report of project undertaken at CEDAR, University of Reading.

Chalupa W, Boston R, Sniffen C J and Miner W H (1999) Formulating rations for dairy cattle on the basis of amino acids; paper to Volac Conference, 18 March, Bugbroke, England.

Chapman H L Jr, Kidder R W and Plank S W (1953) Comparative feeding value of citrus molasses, cane molasses, ground snapped corn and dried citrus pulp for fattening steers on pasture; Florida Agr. Exp. Sta. Bull. 531, 5-16.

Chaucer G (1340-1400) The Miller's Tale; from Canterbury Tales.

Chudy A and Schiemann R (1969) Utilisation of dietary fat for maintenance and fat deposition in model studies with rats. In: Energy Metabolism of Farm Animals, Eds. Blaxter Thorbek and Kielanowski; EAAP Publication No. 12, Oriel Press.

CIRF (1959) Corn Gluten Feed and Gluten Meal and other feeds produced by the corn wet-milling process; Corn Industries Research Foundation Inc, Washington 6, D.C., USA.

Clarke P (1986) Tesco brand cattle cake on way? Farmers Weekly, issue of 5 July 1996.

Close W H (1995) The role of sugar beet feed in the nutrition of pigs: review of recent trials and practical recommendations; report for Trident Feeds, Peterborough, England, July 1995.

COMA (1991) "Fat" in "Dietary Reference Values for Food Energy and nutrients in the United Kingdom", pp 39-60, Department of Health, London, England.

Copeland O C and Shepardson C N (1944) Dried citrus peel and pulp as a feed for lactating cows; Tex. Agric. Exp. Sta. Bull., **658**.

Corporaal J and Harmsen H E (1984) Ensiling and feeding of ensiled maize gluten feed to bulls; Institute for Cattle and Sheep Husbandry, publication No. 27, Lelystad, The Netherlands.

Cole D J A, Beal R M and Luscombe J R (1968) Vet Record **83,** 459-464.

Cox S (1978) Cheaper feed energy from molasses; Farmers Weekly, Supplement to Issue of 8 September, page (iii).

Cranfield H T and Mackintosh J (1935) Taint in milk during the feeding of molassed beet pulp; J. Minist. Agric. (G.B.) **42,** 551-560.

Crawshaw R, Evans R R, Hughes Buddug T and Llewelyn R H (1976) The effects of temperature and the inclusion of formalin on the changes in chemical composition which occur during the storage of liquid whey; In: Liquid whey for dairy cows, ADAS Internal Report.

Crawshaw R (1990) The real value of sugar beet feed; British Sugar Beet Review **58** (3), 42-44.

Crawshaw R (1991) Silage and molassed sugar beet feed; British Sugar Beet Review **59** (4), 23-26.

Crawshaw R (1992 Sheep and sugar beet feed; British Sugar Beet Review **60** (4), 16-19.

Crawshaw R (1992b) Feathermeal: A valuable protein in ruminant rations. Proc. 3rd International Animal Nutrition Symposium, Lisbon, National Renderers Association, Alexandria, VA, USA, 14-18.

Crawshaw R (1994) Bloodmeal: A review of its nutritional qualities for pigs, poultry and ruminant animals. National Renderers Association Technical Review, Alexandria, VA, USA.

Crawshaw R (1994b) Feedfat for pigs; Feed Compounder, **14 No. 7**, August 1994, 25-32.

Crawshaw R (1995) Meat and Bone Meal: A review of its nutritional qualities for pigs. Feed Compounder, **15, No. 7**, August 1995, 14-19.

Crawshaw R, Thorne D M and Llewelyn R H (1980) The effects of formic and propionic acids on the aerobic deterioration of grass silage in laboratory units. J. Sci. Food and Agriculture 31, 685-694.

Cuddeford D (1996) Equine Nutrition; D & N Publishing, Marlborough, England, page 21.

Cunha T J, Pearson A M, Glasscock R S, Bushman D M and Folks S J (1950) Preliminary observations on the feeding value of citrus and cane molasses for swine; Florida Agr. Exp. Sta. Circ. S-10.

Dawkins C W C and Meadowcrof S C (1962) Feeding of brewers' yeast to dairy cows; Experimental Husbandry No. 8, pp 49-55.

Deaville E R, Smith N, Hawes W S and Whitelam (2000) The effect of commercial-scale processing on the degree of .... in animal feeds. In: Recent Advances on Animal Nutrition 2000, pp 56-70, edited by Garnsworthy and Wiseman, Nottingham University Press, Nottingham, England.

Dickson I A and Laird R (1976) Dried sugar beet pulp as a roughage substitute for pregnant ewes; Anim. Prod. **22,** 115-121.

DOE 1 (1993) Communication from The Lord Strathclyde, Department of the Environment to The Malt Distillers Association, 19 May 1993.

DOE 2 (1993) Communication from the Minister for the Environment and Countryside, David MacLean MP to Mr D Padgett of Argrain 24 May 1993.

DOH (1994) Nutritional Aspects of Cardiovascular Disease; Department of Health Report No 46 by the Cardiovascular Review Group Committee on Medical Aspects of Food Policy.

Driggers J C, Davis G K and Mehrhof N R (1951) Toxic factor in citrus seed meal; Florida Agr. Exp. Sta. Tech. Bull.476, 5-36.

EC (1990) Draft list of ingredients for compound feedingstuffs; working document of an EC Mini-Group, 4979/VI/90EN.

EEC (1983) Council Directive 83/228/EEC of 18 April 1983 on the fixing of guidelines for the assessment of certain products used in animal nutrition.

Edwards N J and Parker W J (1995) Apple pomace as a supplement to pasture for dairy cows in late lactation; Proc. NZ Soc. Anim. Prod. **55,** 67-69.

Edwards Sandra A (1982) Feeding potatoes to intensive beef; 1981 Annual Review, ADAS Boxworth Experimental Husbandry Farm, Cambridgeshire, England, pp 44-46.

Edwards Sandra A (1993) Determination of digestible energy of liquid potato feed for pigs; SAC Contract report No 93/204P for James & Son (Grain Merchants) Ltd, Northampton, England.

Edwards Sandra A and Livingstone R M (1990) Potato and potato products; In: Nontraditional Feed Sources for Use in Swine Production, Eds. Thacker and Kirkwood, Butterworths, London, Chapter 31, pp 305-314.

El Hag G A and Miller T B (1972) Evaluation of whisky distillery by-products: IV; The reduction in digestibility of malt distillers' grains by fatty acids and the interaction with calcium and other reversal agents; Journal of the Science of Food and Agriculture, **23**: 247-258.

EPA (1994) International anthropogenic methane emissions: Estimates for 1990; United States Environmental Protection Agency Report to Congress, EPA 230-R-93-010, edited by M J Adler, January 1994.

EU (2000) Draft Commission Decision amending Commission Decision 91/516/EEC establishing a list of ingredients whose use is prohibited in compound feedingstuffs; SANCO 2724/99.

EU (2001) Draft Commission Regulation establishing a list of materials whose circulation or use for animal nutrition purposes is restricted or prohibited.

Eurolysine (1988) Apparent ileal digestibility of essential amino acids in feedstuffs for pigs: Eurolysine Information No. 15, November 1988, Paris, France.

Ewing W N (1997) Corn Steep Liquor; in The Feeds Directory, published by Context, Heather, England, page 22.

Fairbairn C B (1974) Feeding sugar beet pulp to farm livestock; ADAS Q. Review **15,** 114-122.

F.A.O. (2000) Citrus production; Web-site of the Food and Agriculture Organisation, United Nations.

Fegeros K, Zervas G, Stamouli S and Apostolaki E (1995) Nutritive value of dried citrus pulp and its effect on milk yield and milk composition of lactating ewes; J.Dairy Sci. **78.** 1116-1121.

Firkins J L, Berger L, Fahey G C Jr and Merchen N R (1984) Ruminal nitrogen degradability and escape of wet and dry distillers grains and wet and dry corn gluten feeds; J.Dairy Sci. **67;** 1936-1944.

Firkins J L, Berger L and Fahey G C Jr (1985) Evaluation of wet and dry distillers grains and wet and dry corn gluten feeds for ruminants; J. Anim. Sci. **60:3,** 847-860.

Flour Advisory Bureau Our Choice of Bread leaflet, Flour Advisory Bureau Ltd, London, England.

FOB Doorstep to today's loaf; extract by Food and Drink Federation, London, England from Federation of Bakers leaflet.

Fontenot J P, Bovard K P, Oltjen R R, Rumsey T S and Priode B M (1977) Supplementation of apple pomace with non-protein nitrogen for gestating beef cattle, I-Feed intake and performance; J. Anim. Sci. **46,** 513-522.

Fox PF and O'Connor F (1969) Ir. J. agric. Res. **8,** 183.

Freedom Food: leading the way in animal welfare; RSPCA assurance of farm animal welfare; Freedom Food Ltd, Horsham, England RH12 1HG.

Freedom Food (2000) Rapid progress towards full farm assurance status; Focus – Freedom Food News Issue 15, Winter 2000.

FSA (2000) Review of BSE Controls; Report of the Food Standards Agency, London, England, December 2000.

Gadient M and Wegger Inger (1985) Ascorbic Acid in Intensive Animal Husbandry; Roche Information Service, translated and adapted from Annual Report of the Sterility Research Institute, Royal Veterinary and Agricultural University. Copenhagen 1984, **27,** 1-44.

Gasa J, Castrillo C, Guada J A and Balcells (1992) Rumen digestion of ensiled apple pomace in sheep: effect of proportion in diet and source of nitrogen supplementation; Anim. Feed Sci. Technol. **39,** 193-207.

Gerritse R G and Zugec (1977) J. Agric. Sci. (Camb.) **88,** 101-109.

Gillespie Fiona (1985) Molasses increases dry matter intake; Farm Mixer, October/November issue, page 18.

Givens D I and Barber W P (1987) Nutritive value of apple pomace for ruminants; Anim. Feed Sci. Technol. **16,** 311-315.

Givens D I, Moss Angela R and Everington Jeannie M (1992) Nutritional value of cane molasses in diets of grass silage and concentrates fed to sheep; Animal Feed Science and Technology **38,** 281-291.

Gizzi Giselle (2001) Distillers dark grains in ruminant nutrition; Nutr. Abs. & Rev. (in press).

Glasscock R S, Cunha T J, Pearson A M, Pace J E and Buschman D M (1950) Preliminary observations on citrus seed meal as a protein supplement for fattening steers and swine; Florida Agr. Exp. Sta. Circ. S-12.

Glover B (1997) The world encyclopedia of beer, published by Lorenz Books, London.

Gohl B I (1973) Citrus by-products for animal feed; World Animal Review No. 6, pp 24-27.

Graham H (1988) ISI Atlas of Science: Dietary fibre concentration and assimilation in swine; Animal and Plant Sciences 76-80.

Grieve D G and Burgess T D (1977) Beef Industry Research Report, University of Guelph.

Griffiths J G and Crawshaw R (1977) Feeding whey to dairy cows: report of a field study; Internal ADAS report.

GVA (2001) The production of vodka; Website of the Gin & Vodka Association, 20 April 2001, www.ginvodka.org.

Hall R W (1995) The development of the brewery co-products market; The Brewer, April 1995.

Hanrahan T J (1969) Lagoon-stored whey in the diet of growing-finishing pigs; Ir. J. agric. Res. **8,** 271-277.

Hanrahan T J (1977) Proc. Pig Health Soc., 5th Winter Symposium.

Harborne J B (1999) An overview of antinutritional factors in higher plants; In: Secondary Plant Products, Eds. Caygill and Mueller-Harvey. Nottingham University Press, pp 7-16.

Harland Janice (1981) Pressed sugar beet pulp; British Sugar Beet Review; **49 (3),** 47-49.

Harland Janice (1985) A new approach to feeding horses; British Sugar Beet Review **53 (2),** 20-22.

Hemingway R G and Parkins J J (1972) A molassed sugar beet nut containing added urea, phosphate, trace elements and vitamins; British Sugar Beet Review **40 (5),** 207-214.

Hemingway R G, Parkins J J, Ritchie N S, Fishwick G and Fraser J (1976) Urea-containing molassed sugar beet pulp products as major components of barley based diets for growing Friesian steers; Exp. Husb. **30,** 68-73.

Hendrickson R and Kesterson J W (1965) By-products of Florida Citrus: Composition, Technology and Utilisation; Agricultural Experiment Stations, Institute of Food and Agricultural Sciences, University of Florida, Bulletin 698.

Holness and Mandisodza (1985) The influence of additional fat in the diet of sows before and after parturition on piglet viability and performance; Livestock Prod. Sci. **13,** 191-198.

Hulme A C and Rhodes J C (1971) Pome Fruits. In: Biochemistry of Fruits and their Products, edited by A C Hulme, Academic Press, Chapter 10, pp 333-373.

Hutton K (1987) Citrus pulp in formulated diets; Recent Advances in Animal Nutrition in Australia, 1987, Ed. D J Farrell, pp297-316.

Hutton T (2001) Food Manufacturing: An Overview; Key Topics in Food Science and Technology – No. 3, published by Campden & Chorleywood Food Research Association Group, Chipping Campden, England.

Hyslop J J (1991) The storage and nutritional value of wet malt distillers grains for ruminants; Ph.D thesis, University of Glasgow.

Hyslop J J and Roberts D J (1989) Effects of replacing barley/soya with malt distillers grains (draff) in silage based complete diets for dairy cows; Anim. Prod. **48,** 636 Abstr.

Hyslop J J, Offer N W and Barber G D (1989) Effect of ensilage method on storage dry matter loss and feeding val;ue of malt distillers grains (draff); Anim. Prod. **48:3,** 664 Abstr.

IOB (1997) Modified IOB mashing procedure. In: Institute of Brewing Methods of Analysis Volume 1 Analytical; 2.3 Hot water extract of ale, lager and distilling malts, 4.6 Hot water extract of untreated grain grits and other raw grains.

Jenkins D J A, Wolever T M S, Leeds A R, Gassull M A, Haisman P, Dilawari J, Goff D V, Metz G L and Alberti K G MM (1978); British Medical Journal, **1,** 1392-1394.

James & Son (1999) Greenwich Gold; promotional leaflet, James & Son (Grain Merchants Ltd), Northampton, England.

Johnstone C (1991) Preservation of distillery by-products: a compilation of the results of investigations into finding suitable preservatives for draff, Curne Gold, Supergrains and pot ale syrup; unpublished report by United Distillers' Biotechnology Group, United Distillers (UK) Ltd, Edinburgh, Scotland.

Jones I J R (1992) Practical experiences of the Meura 2001Mash Filter at Guinness Park Royal, Brewers' Guardian, October 1992, pp 21-24.

Jongbloed A W, Everts H and Kemme P A (1991) Phosphorus availability and requirements in pigs. In: Recent Advances in Animal Nutrition – 1991, Eds. Haresign and Cole, Butterworths, London, England, Chapter 4, pp 65-80.

Joshi V K and Sandhu D K (1996) Preparation and evaluation of an animal feed byproduct produced by solid-state fermentation of apple pomace; Bioresource-Technology **56: 2-3,** 251-255.

Just A, Fernandez J A and Jorgensen H (1983) Livestock Production Science **10,** 171-186.

Kay R M and Simmins P H (1990) Research Reports from ADAS Terrington for British Sugar plc, Peterborough, England.

Karalazos A and Swan H (1976) Molasses and its by-products. In: Feed Energy Sources for Livestock; edited by Swan and Lewis, Butterworths, London, England, pp 29-46.

Kellner O (1908) The Scientific Feeding of Animals, English translation by W. Goodwin (1926), Duckworth, London, England, p136.

Kennedy A (1987) Determination of water in pot ale syrups; dissertation presented in part fulfilment of of requirements of M.Sc degree, University of Aberdeen.

Kent-Jones D W and Amos A J (1967) Modern Cereal Chemistry, Food Trade Press Ltd.

Kesterson J W (1961) A discussion of the methods for the production of essential oils in Florida. In: Transactions of the 1961 Citrus Engineering Conference; Florida Section A.S.M.E., Lakeland, Florida, USA.

Kirk W G, Kelly E M, Fulford H J and Henderson H E (1956) Feeding value of citrus and blackstrap molasses for fattening cattle; Florida Agr. Exp. Sta. Bull. 575, 3-23.

KW (2000) KW Bakery Products: Processed Bread; Information Sheet, KW Alternative Feeds, Wetherby, England.

Lamming G E, Thomas P C, Maclean C and Cooke E M (1992) The Report of the Expert Group on Animal Feedingstuffs, HMSO.

Lea A.G.H. (1989) Cider Vinegar. In: Processed Apple Products, edited by D L Downing D.L. AVI Van Nostrand, New York.

Lee Pauline A. and Crawshaw R (1991) Molassed sugar beet feed for pigs; British Sugar Beet Review **59** (2), 57-60.

Lewicki W (1997) Cane molasses and its use in the fermentation industry in special view of the process of yeast production; In: Proceedings of the VH-Yeast Conference, Vienna, Austria; pp 55-66.

Lewis M and Lowman B G (1989) Ensiled distillers or brewers' grains as the sole diet for beef cattle; Animal Production, **48 (3):** 656 Abstr.

Lewis M, Ramsay S, Offer N W and Sinclair K (1990) An appraisal of the relative value of malt distillers grains purchased in the summer for ensiling; SAC Report prepared for United Distillers plc, Edinburgh, Scotland.

Lewis M and Scott N A (1990) Nutritive value of Supergrains and its use as a feed for finishing beef cattle; IRC Biotechnology Group External Research Report No. 1990/16 on behalf of United Distillers, plc, Edinburgh, Scotland.

Ling E R (1956) A Textbook of Dairy Chemistry; 3rd Edition Revised, Volume 1, published by Chapman & Hall, London.

Lloyd W J W (1986) Adjuncts; J. Inst. Brew. **92,** July-August , 336-345.

Lofgreen G P and Otagaki K K (1960) J. Anim. Sci. **19,** 392.

Longland Annette C, Close W H and Low A G (1990) Research Report from AFRC Shinfield for British Sugar plc, Peterborough, England.

Lonsdale C R (1989) Straights: Raw Materials for Animal Feed Compounders and Farmers; published by Chalcombe Publications. Address at time of publication was Marlow, England but current address is Lincoln, England.

Losada H and Preston T R (1974) Cuban J. Agric. Sci. **8,** 11.

Lucas I A M (1966) Pig Farming, May issue, p 70.

McBurney M I, van Soest P J and Chase L E (1983) J. Sci. Food and Agric. **34,** 910.

McCance R A and Widdowson Elsie M (1991) Composition of Foods; 5th Edition, edited by Holland, Welch, Unwin, Buss, Paul and Southgate, published by Royal Society of Chemistry, Cambridge and the Ministry of Agriculture, Fisheries and Food, London, England.

McDonald P (1981) The Biochemistry of Silage; John Wiley & Sons.

McGee H (1984) On Food and Cooking; published by Unwin Hyman Ltd, London.

McKendrick Elizabeth J and Hyslop J J (1991) The evaluation of Supergrains as a concentrate feed for dairy cows; IRC Biotechnology Group External Research Report No. 1991/13 on behalf of United Distillers plc, Edinburgh, Scotland.

McKendrick Elizabeth J and Hyslop J J (1992) A comparison of distillers dark grains with proprietary concentrate for milk production; Animal Production, Paper No. 62.

MAFF (1961) Manual of Nutrition, Sixth Edition, HMSO.

MAFF (1976) Technical Bulletin 33: Energy Allowances and Feeding Systems for Ruminants, HMSO.

MAFF (1983) Mineral, Trace Element and Vitamin Allowances for Ruminant Livestock: The report of an interdepartmental Working Party set up to consider the findings contained in the ARC (1980) Technical Review – The Nutrient Requirements of Ruminant Livestock, page 12.

MAFF (1990) UK Tables of Nutritive Value and Chemical Composition of Feedingstuffs, Eds. Givens, Hopkins, Morgan, Stranks, Topps and Wiseman, published by Rowett Research Services Ltd, Aberdeen, Scotland.

MAFF (1991 and 1998) Code of good agricultural practice for the protection of water, MAFF Publications, London.

MAFF (1993) Prediction of the Energy Values of Compound Feeding Stuffs for Farm Animals; Summary of the recommendations of a working party sponsored by the Ministry of Agriculture, Fisheries and Food.

MAFF (1995) Code of Practice for the control of salmonella; during the storage, handling and transport of raw materials intended for incorporation into, or direct use as, animal feedingstuffs; MAFF Publications, London.

MAFF (1998) National Food Survey, p13.

MAFF (2000) The Feeding Stuffs Regulations 2000, HMSO.

MAFF (2001) Retail Production of Animal Feedingstuffs in Great Britain, December 2000 to February 2001; MAFF National Statistics.

Malting and Brewing Science (1982) Methods of wort boiling and hop extraction; Volume 2, edited by Hough J S, Briggs D E, Stevens R and Young T W, published by Chapman and Hall, Chapter 15, page 514 .

Manley D (1998) Biscuit, Cookie and Cracker Manufacturing Manual; Woodhead Publishing Ltd.

Manterola B H, Cerda A D, Mira J J, Porte F E, Luis-Sirhan A and Casanova G (1999) Effect of including high levels of apple pomace on dry matter and crude protein degradability and ruminal and blood parameters, Avances en Produccion Animal, **24: 1-2,** 31-39.

Martin P A (1982) Calculation of calorific value of beer; J. Inst. Brew. September-October 1982, **88,** 320-321.

Marty R J and Preston T R (1970) Rev. cubana Cienc. Agric. (English edition) **4,** 183.

MDA (1994) The Distilling Equation; The Malt Distillers' Bulletin, Spring 1994, published by The Malt Distillers Association of Scotland.

MMF Speedi-beet: Quick soaking sugar beet flakes, leaflet produced by Masham Micronised Feeds, Ripon, England.

Megias M D, Cherney J H and Cherney D J R (1997) Effects of phenolic compounds in cell walls of orange and artichoke by-product silages on in vitro digestibility; J. Appl. Anim. Res. **12,** 127-136.

Miller T B (1969) Evaluation of whisky distillery by-products. I. Chemical composition and losses during transport and storage of malt distillers' grains; J. Sci. Food Agric. **20:** 477-480.

Mitchell L M (1990) Dry matter digestibility of intensive beef diets based on distillery by-products; IRC Biotechnology Group External Research Report No. 1990/13 on behalf of United Distillers and Pentland Scotch Whisky Research Ltd, Edinburgh, Scotland.

Mitchell K G and Sedgewick P H (1963) The effect on the performance of growing pigs of the level of meal fed in conjunction with an unrestricted supply of whey; J. Dairy Res. **30,** 35-45.

Morgan E K, Gibson M L, Nelson M L and Males J R (1991) Utilisation of whole or steamrolled barley fed with forages to wethers and cattle; Animal Feed Science and Technology **33,** 59-78.

Moser B D, Boyd D and Cast W R (1978) Piglet survival; Nebraska Swine Report, Nebraska, USA, p4.

Mueller-Harvey Irene (1999) Tannins: their nature and biological significance. In: Secondary Plant Products, edited by Caygill and Mueller-Harvey, Nottingham Univ. Press, chapter 3, pp 17-39.

NABIM (1998) Background note on pesticide residues, National Association of British and Irish Millers, London, England.

NACM (1992) Code of Practice for the Production of Cider and Perry, National Association of Cider Makers, London, England.

NACM (2000) Website of the National Association of Cider Makers, London, England.

NACNE (1983) A discussion paper on proposals for nutritional guidelines for health education in Britain; a publication prepared by an ad hoc worhing party under the chairmanship of Prof. W P T James for The Health Education Council, Department of Health and Social Security, London, England.

NDC (1999) Dairy Facts & Figures 1999 edition; National Dairy Council, London, England.

NDC (2000) The nutritional value of milk; In: Milk Matters – a Guide to the UK Dairy Industry and its Products; a publication of the National Dairy Council, London, England.

NDFAS (1998a) National Dairy Farm Assured Scheme: Standards document available from the NDFAS Office, Reading, England.

NDFAS (1998b) Code of Practice for the safe storage, handling and feeding of feed materials on farm; available from the National Dairy Farm Assured Scheme Office, Reading, England.

Neame C (1981) New use for brewers' yeast; Feed Manufacture, February 1981.

Nielsen W K (1992) Membrane filtration; Marketing Bulletin of APV Pasilac AS, Denmark, edited by Per Nielsen, Third revised edition, APV, Aarhus, Denmark.

Norfolk Agricultural Station (1981) Feeding pressed beet pulp to beef cattle: preliminary results; Information sheet 81/10, Norfolk, England.

Novartis Nutrition Introducing Malt Matters; internal document – Novartis Consumer Health UK Ltd, Kings Langley, England.

NRC (1989) Nutrient requirements of horses, 5th revised edition, National Academy Press, Washington D.C., USA.

NRC (1998) Nutrient requirements of swine, 10th revised edition, National Academy Press, Washington D.C., USA.

Offer N W (1992) The protein value of dried distillers grains. Final Report to Pentlands Scotch Whisky Research Ltd, Edinburdh, Scotland, 16 pp.

Offer N W (1999) Fresh citrus pulp for dairy cows; Report to James & Son (Grain Merchants) Ltd, Northampton, England.

Offer N W and Al-Rwidah M N (1989) The use of absorbent materials to control effluent flow from grass silage: experiments with drum silos; Research and Development in Agriculture **6 (2),** 71-76.

OJ No L 78/32-37 (1991) Council Directive of 18 March amending Directive 75/442/EEC on waste (91/156/EEC); Official Journal of the European Communities.

Orchard House Foods (1997) Code of Practice: For the handling of citrus fruit destined for freshly squeezed juice, Corby, England.

Owen F G and Larson L L (1991) Corn distillers dried grains versus soybean meal in lactation diets; J. Dairy Sci. **74,** 972-979.

Oxford (1995) The Concise Oxford Dictionary; Ninth Edition edited by Della Thompson, published by BCA, p 1060.

Palmquist D L (1984) Use of fats in diets for lactating dairy cows; In: Fats in Animal Nutrition, edited by J Wiseman, published by Butterworths, Sevenoaks, England, Chapter 18, pp 357-381.

Parkins J J, Hemingway R G and Ritchie N S (1974) Molassed sugar-beet nuts supplemented with urea and phosphate used as a milk production concentrate for dairy cows; J. Dairy Research **41,** 289-297.

Pascual J M and Carmona J F (1980a) Composition of citrus pulp; Animal Feed Science and Technology **5,** 1-10.

Pascual J M and Carmona J F (1980b) Citrus pulp in diets for fattening lambs; Animal Feed Science and Technology **5,** 11-22.

Pass R T and Lambart I (2001) Co-products**;** In: Handbook of Alcoholic Beverages, Volume 2 Whisky, edited by G Stewart, published by Academic Press (In Press).

Pass R T, Dewey P J S and Livingstone R M (1989) The metabolizable energy contents of distillery by-products. In: Proceedings of the 50th Easter School of Agricultural Science, Nottingham University.

Penrose J D F (1982) Upgrading grains; The Brewer, January 1982, pp4-7.

Perrott G (1993) Grainbeet – a valuable feed for ruminants; British Sugar Beet Review **60: No 1,** 39-42.

Pettigrew J E (1981) Supplemental dietary fat for peripartal sows, A Review; J. Anim. Sci. **53:1**, 107-117.

Pettigrew J E and Moser R L (1991) Fat in swine nutrition. In: Swine Nutrition, Eds. Miller, Ullrey and Lewis, Butterworth Heinemann, pp133-140.

Pickles (1968) Chorleywood Bread Process, UK Patent No. 1133472, Campden and Chorleywood Food Research Association Group, Chipping Campden, England.

Pilnik W and Voragen A G J (1970) Pectic substances and other uronides; In: Biochemistry of Fruits and their Products, edited by A C Hulme, chapter 3, pp 53-87.

Pinzon F J and Wing J M (1975) Effects of citrus pulp in high urea rations for steers; J. Dairy Science **59,** 100-1103.

Pollard A and Timberlake C F (1971) Fruit Juices. In:Biochemistry of Fruits and their Products, Ed. A C Hulme, chapter 17, pp 573-621.

Randall R P, Wallenius R W, Fyer I A and Hillers J K (1972) Use of molassed dried beet pulp–urea as an NPN source for young ruminants; J.Anim. Sci. **35,** 1083-1086.

Ratcliff J (2000) Supermarkets: powerful consumer advocates; Feed Tech **4, No.5/ 6,** 30-31.

Rhône-Poulenc (1997A) Use of enzymes in alcohol production; Applications Bulletin of Rhône-Poulenc ABM Brewing & Enzymes Group, Stockport, England.

Rhône-Poulenc (1997B) BG 200; Applications Bulletin of Rhône-Poulenc ABM Brewing & Enzymes Group, Stockport, England.

Roth-Maier D A (1979) Die verfutterung von bierhefe ist vielseitiger geworden; Tierzuchter **31 (4),** 159-160.

Rooke J A, Moss Angela R, Mathers A I and Crawshaw R (1997) Assessment using sheep of the nutritive value of liquid potato feed and partially fried potato chips (french fries); Animal Feed Science and Technology **64,** 243-256.

Rooke J A (1999) Evaluation of a potato product for newly-weaned pigs; SAC Experimental Report on behalf of James & Son (Grain Merchants) Ltd, Northampton, England.

Rowett Research Institute (1983) Annual Report of Studies in Animal Nutrition and Allied Sciences; **39,** 93.

Rowett Research Institute (1984a) Malt Distillers' draff, and mixture of draff and pot ale syrup; in Fourth report of Feedingstuffs Evaluation Unit, Aberdeen, Scotland, pp 65-67.

Rowett Research Institute (1984b) Brewers' grains; in Fourth report of Feedingstuffs Evaluation Unit, Aberdeen, Scotland, pp 73-77.

Rowett Research Institute (1984c) Pot ale syrup**;** in Fourth report of Feedingstuffs Evaluation Unit, Aberdeen, Scotland, pp 68-70.

Rowett Research Institute (1984d) Distillers' Dark Grains**;** in Fourth report of Feedingstuffs Evaluation Unit, Aberdeen, Scotland, pp 61-64.

Rowett Research Institute (1985) Annual Report of Studies in Animal Nutrition and Allied Sciences, RRI, Aberdeen, Scotland, **40,** 103.

Rowett Research Institute (1990) Enzyme supplementation of grain-based and other diets for non-ruminants; Final report of link project, RRI, Aberdeen, Scotland.

Rumsey T S and Lindahl I L (1982) Apple pomace and urea for gestating ewes; J. Anim. Sci. **54,** 221-234.

Rusznyak I and Szent-Gyorgyi A (1936) Vitamin P: flavanols and vitamins; Nature **138,** 27.

Rymer Caroline (1988) Digestion of molassed sugar beet pulp by ruminants; Ph. D thesis, University of Newcastle upon Tyne.

SAC (1997) The Farm Management Handbook 1997/98, 18th edition edited by Linda Chadwick, Scottish Agricultural College, Edinburgh, Scotland.

Savery C R (1984) Maize gluten feed; Nutrition Chemistry Technical Bulletin 84/ 106, ADAS South East Region.

Schingoethe D J (1976) Whey utilisation in animal feeding: a summary and evaluation; J. Dairy Sci. **59: 3,** 556-570.

Schwab C G and Satter L D (1976) J. Dairy Sci **59;** 1254.

Scottish Office (1993) Communication from The Secretary of State, Mr Ian Lang MP, to Dr A Rutherford, Malt Distillers Association, 11 March 1993.

Seerley R W (1984) The use of fat in sow diets. In: Fats in Animal Nutrition; Ed Wiseman, 333-352, Butterworths, Sevenoaks, England.

Shenstone W A (1895) Justus von Liebig: His Life and Work (1803-1873), Macmillan, New York.

Stanhope D L, Hinman D D Everson D O and Bull R C (1980) Digestibility of potato solanum-tuberosum processing residue in beef cattle finishing diets; J.Anim. Sci. **51 (1),** 202-206.

Steckley J D, Grieve D G, MacLeod G K and Moran Jr. E T (1979) Brewers' yeast slurry 1. Composition as affected by length of storage, temperature and chemical treatment; J. Dairy Sci. **62,** 941-946.

Suttle N F (1999) Minerals in Animal Nutrition; published by CAB International, pp 322-325.

Suttle N F, Brebner J and Pass R T (1996) A comparison of the availability of copper in four whisky distillery by-products with that in copper sulphate for lambs; British Society of Animal Science **62,** 689-690.

Sutton J D, Daley S R, Haines M J and Thomson D J (1988) Comparison of dried molassed and unmolassed sugar beet feed and barley at two protein levels for milk production in early lactation; Anim. Prod. **46,** 490 Abstr.

SWA The World of Scotch Whisky: Spirit of Scotland; published by the Scotch Whisky Association, Edinburgh, Scotland.

SWA (1997) Scotch Whisky: Questions and Answers; published by the Scotch Whisky Association, Edinburgh, Scotland, p 9.

SWA (1999) Statistical Report of the Scotch Whisky Association, Edinburgh, Scotland, page 3.

Swan H (1978) The liquor for all livestock; Farmers Weekly, Supplement to Issue of 8 September, page (iii).

Thomas P C, Robertson S, Chamberlain D G, Livingstone R M, Garthwaite P H, Dewey P J S, Smart R and White C (1989) Predicting the metabolisable energy (ME) content of compound feeds for ruminants; In: Recent Advances in Animal Nutrition 1988, Eds Haresign and Cole, Butterworths, London, pp 127-146.

Thomas E (2000) The basics of biotechnology; Bulletin of the National Renderers Association, Issue Number 823, pp 4-5.

Thomas S (1996) The use of partially fried chips in diets for intensive bull beef; an SAC report for James & Son (Grain Merchants) Ltd, Northampton, England.

Thonney M L, Fox D G, Duhaine D J and Hoque D E (1986) Fish meal and apple pomace for growing cattle; Proc. Cornell Nutrition Conference for Feed Manufacturers, pp 106-114.

Ting S V and Attaway J A (1971) Citrus fruits. In: The Biochemistry of Fruits and their Products, Volume 2, Ed. A C Hulme, Academic Press, p 127-128.

Todd N (1977) Whey bloat in fatteners; Pig Farming Supplement, October issue, pages 51 and 55.

Trident Feeds I Distillers Feeds: A guide to using distillers feeds on the farm, Trident Feeds, Peterborough, England.

Trident Feeds Grainbeet Handbook: A guide to using the Grainbeet system on the farm, Trident Feeds, Peterborough, England.

Trident Feeds FF3 Pressed Sugar Beet Pulp; Feed Facts – Product Bulletin from Trident Feeds, Trident Feeds, Peterborough, England.

UKASTA (1997 and subsequent versions in 1998, 1999 and 2000) Code of Practice for Road Haulage (of combinable crops, animal feed materials and as-grown seeds); produced by the United Kingdom Agricultural Supply Trade Association Ltd, London, England.

UKASTA Code of Safe Practice for the Storage, Packaging and Supply of Animal Feed Materials which are supplied to farms; produced by the United Kingdom Agricultural Supply Trade Association Ltd, London, England.

UKASTA (2001) European 'Yes' to UK Feed Assurance Scheme; Press Release of 23 February, United Kingdom Agricultural Supply Trade Association Ltd, London, England.

Unigate (1997) The Unigate Business Deal: Superior Stockmanship Standards; The Unigate Agricultural Affairs Department, Unigate European Food, Swindon, England.

Es A J H van, Kijkamp H J and Vogt J E (1971) The net energy content of dried sugar-beet pulp and of sucrose when fed to lactating cows; Neth. J. Agric. Sci. **19,** 48-56.

Van Soest P J, McBurney M I and Russell J (1984) Proc. of California Animal Nutrition Conference, p53, Pomona, Ca., USA.

Van Soest P J and Mason V C (1991) The influence of the Maillard reaction upon the nutritive value of fibrous feeds. Animal Feed Science and Technology, **32:** 45-53.

Varel V H, Pond W G, and Yen J (1984) Influence of dietary fibre on the performance and cellulase activity of growing swine; J. Anim. Sci. **59** (2), 388-393.

Varel V H (1987) Activity of fibre-degrading micro-organisms in the pig large intestine; J. Anim. Sci. **65,** 488-496.

Versteegh H A J, Jongbloed A W and van Diepen J Th M (2000) The nutritive value of the wet by-product wheat bottom stills (Greenwich Gold) for pigs; report to Amylum Group, Aalst, Belgium.

VITEC 3a (1988-1990) The effects of feed ingredients on egg yolk pigmentation;

Roche Animal Nutrition and Vitamin News C2-1/1, Roche Products Ltd, Welwyn Garden City, England.

VITEC 3b (1988-1990) Maize germ meal; Roche Animal Nutrition and Vitamin News G2 – 12/1, Roche Products Ltd, Welwyn Garden City, England.

VITEC-3c (1988-1990) Biscuit Meal; Roche Animal Nutrition and Vitamin News G2 – 4/1, Roche Products Ltd, Welwyn Garden City, England.

Wackerbauer K, Zufall C and Holscer K (1993) The influence of grist from a hammermill on wort and beer quality, Brauwelt International, No2, pp 107-113.

Wainman F W and Dewey P J S (1988) Maize by-products of UK origin; In: Feedingstuffs Evaluation Unit, Fifth Report 1988, Rowett Research Institute, Aberdeen, Scotland, pp 16-34.

Wainman F W, Dewey P J S and Brewer A C (1985) The ME values of mixtures of malt distillers draff and barley flour with particular attention to methane energy losses; unpublished report of work conducted at the Rowett Research Institute, Aberdeen, Scotland.

Walker A J (1983) The suitability of brewery byproducts as feeds for finishing pigs; ADAS Internal Report.

Walker N (1985) Molasses for in-pig sows and finishing pigs; Irish Farmers' Journal, Issue of 16 November, page 16.

Wander I Extra Malt leaflet, Wander Ltd (now Novartis Nutrition UK), Kings Langley, England.

Wander II Scientific proof of the remarkable value of Ovaltine for nervous children; promotional leaflet by Wander Ltd (now Novartis Nutrition UK), Kings Langley, England.

Webster A J F and Chaudhry A S (1992) Characterisation of protein in animal feeds with particular reference to distillers' grains; Final report to Pentlands Scotch Whisky Research Ltd, Edinburgh, Scotland, 44pp.

Welch J G, Nilson K M and Smith A M (1973) Feeding liquid whey to dairy cattle; J. Dairy Sci. **57,** 634 Abstr.

West E S and Todd W R (1957) Textbook of Biochemistry, Second Edition, Macmillan, New York, pp 250-251.

Whittemore C T, Taylor A G and Elsley F W H (1973) The influence of processing upon the nutritive value of the potato: Digestibility studies with pigs, J. Sci. Fd. Agric. **24,** 539-545.

Whittemore C T, Taylor A G, Moffat I W and Scott A (1975) Nutritive value of raw potato for pigs, J. Sci. Fd. Agric. **26,** 255-260.

Whyte A (1993) Faecal digestibility trial to evaluate the apparent and true digestibility of dark distillers grains in rats; Report by the Rowett Research Services Ltd, Aberdeen, Scotland.

Wilkinson J M and Kendall N (1997) The value of liquid potato feed for dairy cows; report by Chalcombe Agricultural Resources to James & Son (Grain Merchants) Ltd, Northampton, England.

Willard M J, Hix V M and Kluge G (1987); In: Potato Processing, 6th Edition, Ed. Talbot and Smith, published by AVI, pp 566-581

Wilmot K (1988) When the chips are down, Processing June 1988, pp 21-27.

Woodman H E and Calton W E (1928) The composition and nutritive value of sugar beet pulp; J. Agic. Sci. Camb. **18,** 544-568.

WHO (1982) Prevention of Coronary Heart Disease, World Health Organisation Expert Committee Technical Report Series **678.**

WPSA (1992) European Amino Acid Table; 1st edition, published by World Poultry Science Association, Beekbergen, The Netherlands.

Wright A (1996) Quality Assurance Scheme; Farmers Weekly, issue of 27 June 1996, p 24.

Wu Z and Huber J T (1994) Relationship between dietary fat supplementation and milk protein concentration in lactating dairy cows: A review. Livestock Production Science **39 No 2,**141-155.

Index

A

B

C

D

E

G

M

N

O

P

Z

ND - #0105 - 050826 - C0 - 234/156/14 - PB - 9781789182743 - Matt Lamination